Daniel Pietsch

Iron-dependent Gene Regulation of a Fresh Water Cyanobacterium

Daniel Pietsch

Iron-dependent Gene Regulation of a Fresh Water Cyanobacterium

Biochemical and Genetic Studies of the Mesophilic Fresh Water Cyanobacterium Synechococcus elongatus PCC 7942

Südwestdeutscher Verlag für Hochschulschriften

Impressum/Imprint (nur für Deutschland/ only for Germany)
Bibliografische Information der Deutschen Nationalbibliothek: Die Deutsche Nationalbibliothek verzeichnet diese Publikation in der Deutschen Nationalbibliografie; detaillierte bibliografische Daten sind im Internet über http://dnb.d-nb.de abrufbar.

Alle in diesem Buch genannten Marken und Produktnamen unterliegen warenzeichen-, marken- oder patentrechtlichem Schutz bzw. sind Warenzeichen oder eingetragene Warenzeichen der jeweiligen Inhaber. Die Wiedergabe von Marken, Produktnamen, Gebrauchsnamen, Handelsnamen, Warenbezeichnungen u.s.w. in diesem Werk berechtigt auch ohne besondere Kennzeichnung nicht zu der Annahme, dass solche Namen im Sinne der Warenzeichen- und Markenschutzgesetzgebung als frei zu betrachten wären und daher von jedermann benutzt werden dürften.

Verlag: Südwestdeutscher Verlag für Hochschulschriften Aktiengesellschaft & Co. KG
Dudweiler Landstr. 99, 66123 Saarbrücken, Deutschland
Telefon +49 681 37 20 271-1, Telefax +49 681 37 20 271-0
Email: info@svh-verlag.de
Zugl.: Bielefeld, Universität, Diss.,2008

Herstellung in Deutschland:
Schaltungsdienst Lange o.H.G., Berlin
Books on Demand GmbH, Norderstedt
Reha GmbH, Saarbrücken
Amazon Distribution GmbH, Leipzig
ISBN: 978-3-8381-1397-5

Imprint (only for USA, GB)
Bibliographic information published by the Deutsche Nationalbibliothek: The Deutsche Nationalbibliothek lists this publication in the Deutsche Nationalbibliografie; detailed bibliographic data are available in the Internet at http://dnb.d-nb.de.

Any brand names and product names mentioned in this book are subject to trademark, brand or patent protection and are trademarks or registered trademarks of their respective holders. The use of brand names, product names, common names, trade names, product descriptions etc. even without a particular marking in this works is in no way to be construed to mean that such names may be regarded as unrestricted in respect of trademark and brand protection legislation and could thus be used by anyone.

Publisher: Südwestdeutscher Verlag für Hochschulschriften Aktiengesellschaft & Co. KG
Dudweiler Landstr. 99, 66123 Saarbrücken, Germany
Phone +49 681 37 20 271-1, Fax +49 681 37 20 271-0
Email: info@svh-verlag.de

Printed in the U.S.A.
Printed in the U.K. by (see last page)
ISBN: 978-3-8381-1397-5

Copyright © 2010 by the author and Südwestdeutscher Verlag für Hochschulschriften Aktiengesellschaft & Co. KG and licensors
All rights reserved. Saarbrücken 2010

Table of Contents

1 **Introduction** 1

   1.1 **Cyanobacteria** 1

      1.1.1 General considerations about cyanobacteria 1
      1.1.2 Morphology of the cyanobacterial cell 1

   1.2 **Cyanobacterial electron transport** 3

      1.2.1 Photosynthetic electron transport 4
          1.2.1.1 Photosynthetic linear electron transport 4
          1.2.1.2 Photosynthetic cyclic electron transport 5
          1.2.1.3 Photosystem I 6
          1.2.1.4 Ferredoxin:NADP$^+$ oxidoreductase 7
      1.2.2 Respiratory electron transport chain 8
      1.2.3 The cyanobacterial NDH-1 complex 9
      1.2.4 L-amino acid oxidase 11
      1.2.5 Interrelationship between photosynthesis and respiration 12

   1.3 **Adaption to iron limitation in cyanobacteria** 14

      1.3.1 The biological use of iron 14
      1.3.2 Interrelationship between iron homeostasis and oxidative stress ... 15
      1.3.3 Consequences of iron limitation in cyanobacteria 17
      1.3.4 Iron acquisition and storage in cyanobacteria 18
      1.3.5 Expression of iron-regulated proteins 20
          1.3.5.1 Characteristics and function of the Isi-protein family 21
          1.3.5.2 Regulation of *isiA* gene expression 23
          1.3.5.3 The IdiA protein 24
          1.3.5.4 Regulation of *idiA* expression 26
      1.3.6 Summary of the modification of PS I by IsiA and PS II by IdiA 29

   1.4 **Aim of research** 31

2 **Materials and Methods** 33

   (A) **General information** 33

   2.1 **Chemicals** 33
   2.2 **Enzymes and kits** 33
   2.3 **Bacterial strains and plasmids** 34

      2.3.1 Bacterial strains 34
      2.3.2 Plasmids 35
      2.3.3 Oligonucleotides 36

I

2.4 Growth conditions and media 38
- 2.4.1 Growth conditions of *S. elongatus* PCC 7942 38
- 2.4.2 Growth of *S. elongatus* PCC 7942 under stress conditions 38
- 2.4.3 Growth of *Escherichia coli* 38
- 2.4.4 Antibiotics used with *S. elongatus* PCC 7942 and *E. coli* 38

(B) Molecular biology methods 39

2.5 Transformation of *E. coli* 39
- 2.5.1 Electroporation of *E. coli* 39
- 2.5.2 Transformation of *E. coli* 39

2.6 Transformation of *S. elongatus* PCC 7942 39

2.7 Isolation of nucleic acids 40
- 2.7.1 Isolation of plasmid DNA from *E. coli* utilizing the Plasmid Miniprep Kit II™ .. 40
- 2.7.2 Isolation of plasmid DNA from *E. coli* with HB-lysis 40
- 2.7.3 Isolation of genomic DNA from *S. elongatus* PCC 7942 40
- 2.7.4 Isolation of total RNA from *S. elongatus* PCC 7942 for Northern blot analysis 40

2.8 Basic genetic methods 41
- 2.8.1 Agarose gel electrophoresis 41
- 2.8.2 DNA restriction digests 41
- 2.8.3 Dephosphorylation of linear DNA fragments 41
- 2.8.4 Phosphorylation of linear DNA fragments 41
- 2.8.5 Ligation of linear DNA fragments 41
- 2.8.6 Fill in of 5`-protruding DNA ends 41
- 2.8.7 Extraction and purification of DNA from agarose gels 41
- 2.8.8 Quantification of DNA and RNA 41
- 2.8.9 RNA integrity and quality test 42

2.9 Polymerase chain reaction 42
- 2.9.1 Polymerase chain reaction using BioTherm™ DNA polymerase 42
- 2.9.2 Polymerase chain reaction using Pwo polymerase 42
- 2.9.3 Polymerase chain reaction using Phusion™ Hot Start DNA- polymerase 42
- 2.9.4 Colony PCR 42
- 2.9.5 Amplification of Dig-dUTP-labelled probes 42

2.10 Purification of DNA samples 43
- 2.10.1 PCR Purification Kit 43
- 2.10.2 Nucleotide Removal Kit 43

2.11 Southern blot 43

2.12 Purification of RNA samples 43

2.13 Hybridisation of RNA 44
2.13.1 Northern blot 44
2.13.2 Slot blot 44
2.13.3 Hybridisation of RNA with Dig-dUTP-labelled probes 44
2.13.4 Colorimetric detection 45
2.13.5 Detection with CDP-Star™ 45

2.14 DNA-microarray experiments 45
2.14.1 Isolation of total RNA from *S. elongatus* PCC 7942 for DNA-microarray experiments 45
2.14.2 Target labelling 45
2.14.3 Hybridisation and wash protocols 46
2.14.4 Image acquisition and data analysis 46

(C) Biochemical methods 47

2.15 Determination of cell growth 47
2.16 Determination of pigment content 47
2.17 Determination of protein content 47
2.18 Preparation of cell suspensions 47
2.19 Preparation of cell-free extracts 48
2.19.1 Preparation of cell-free extracts of *S. elongatus* PCC 7942 with a French Press 48
2.19.2 Preparation of cell-free extracts of *S. elongatus* PCC 7942 using a Ribolyzer . 48
2.19.3 Preparation of cell-free extracts of *E. coli* 48

2.20 Isolation of subcellular fractions 48
2.20.1 Isolation of thylakoid membranes 48
2.20.2 Isolation of periplasmic proteins 49
2.20.3 Isolation of the outer membrane, the cytoplasmic membrane, and the thylakoid membrane 49

2.21 Isolation of PS I complexes from *S. elongatus* PCC 7942 50
2.22 Determination of photosynthetic activity 51
2.22.1 Determination of photosynthetic oxygen evolution 51
2.22.2 Determination of respiratory oxygen uptake 51
2.22.3 Determination of the activity of PS I 51
2.22.4 Determination of the effective quantum yield of PS II 51
2.22.5 77K pigment fluorescence measurements 51

2.23 Recombinant expression of IdiC 52
2.23.1 Heterologous expression of IdiC in *E. coli* 52
2.23.2 Homologous expression of IdiC in *S. elongatus* PCC 7942 52

Table of Contents

- 2.24 Protein purification 53
  - 2.24.1 Solubilisation of recombinant IdiC protein 53
  - 2.24.2 Batch purification of 6 x His-tagged IdiC using Ni-NTA affinity chromatography 53
  - 2.24.3 Preparative SDS PAGE and electroelution 53
  - 2.24.4 Dialysis 53
- 2.25 Generation of a polyclonal antibody against IdiC 54
- 2.26 Production of peptide antibodies against NdhA and NdhB 54
- 2.27 Generation of peptide antibodies against FNR 54
- 2.28 Precipitation of proteins 54
  - 2.28.1 Acetone precipitation 54
  - 2.28.2 Chloroform-methanol precipitation 55
  - 2.28.3 Trichloroacetic acid precipitation 55
- 2.29 Sodium dodecyl sulfate gel electrophoresis (SDS PAGE) 55
  - 2.29.1 Tris glycine SDS PAGE 55
  - 2.29.2 Urea SDS PAGE 55
  - 2.29.3 Tris tricine SDS PAGE 55
- 2.30 Native PAGE 55
- 2.31 2D Blue Native PAGE 56
- 2.32 Staining of protein gels 56
  - 2.32.1 Coomassie Brilliant Blue stain 56
  - 2.32.2 Silver stain 56
- 2.33 Immunoblot analysis 57
  - 2.33.1 Protein transfer 57
  - 2.33.2 Colorimetric detection using 4-chloro-1-naphtol 57
  - 2.33.3 Colorimetric detection using nitroblue tetrazolium and bromo-4-chloro-3-indolylphosphate 57
  - 2.33.4 Chemiluminescent detection 57
  - 2.33.5 Antibodies used in this work 58
- 2.34 Mass spectrometric analysis via MALDI-TOF MS 59
  - 2.34.1 Coomassie Brilliant Blue stain 59
  - 2.34.2 Silver stain 59
  - 2.34.3 Tryptic digest and MALDI-TOF MS 59
- 2.35 Iron determination using ICP-OES 60
- 2.36 Iron determination using EPR 60
- 2.37 IdiC localisation using immunocytochemistry 60

(D) Bioinformatic methods 61

2.38 Database searches and sequence analysis 61

3 Results 62

3.1 Bioinformatic analysis of the idiC gene and its gene product 62

3.2 Expression of IdiC in *S. elongatus* PCC 7942 66
  3.2.1 Expression of IdiC under iron-deficient growth conditions 66
  3.2.2 Expression of IdiC in the late growth phase 68

3.3 Investigation of the localisation of IdiC in *S. elongatus* PCC 7942 70
  3.3.1 Localisation of IdiC in subcellular fractions of *S. elongatus* PCC 7942 70
  3.3.2 Immunocytochemical studies on the localisation of IdiC 72
  3.3.3 Investigation of thylakoid membranes by BN PAGE 72
  3.3.4 Purification and investigation of trimeric PS I complexes 75
    3.3.4.1 Purification of PS I by HIC 75
    3.3.4.2 Second purification step using IEC 76

3.4 Identification of the iron cofactor of IdiC 79
  3.4.1 Homologous expression of 6 x His tagged IdiC in *S. elongatus* PCC 7942 79
  3.4.2 Heterologous expression of 6 x His-tagged IdiC in *E. coli* 83
  3.4.3 ICP-OES-based determination of metal content 84
  3.4.4 Iron determination using EPR measurements 85

3.5 Construction and characterisation of an *idiC*- merodiploid *S. elongatus* PCC 7942 mutant 89
  3.5.1 Attempts to construct an IdiC-free *S. elongatus* PCC 7942 mutant ... 89
  3.5.2 Effect of a reduced IdiC content on growth and photosynthetic activity of *S. elongatus* PCC 7942 WT and the *idiC*-merodiploid mutant MuD 90
  3.5.3 Consequences of reduced IdiC content on the respiratory electron transport activity in the *idiC*-merodiploid *S. elongatus* PCC 7942 mutant MuD 96
  3.5.4 Consequences of reduced IdiC content on the expression of selected iron-regulated genes and proteins in MuD compared to *S. elongatus* PCC 7942 WT 99

3.6 Comparative analysis of growth as well as IdiC, IdiB, and IdiA expression in *S. elongatus* PCC 7942 WT, the *idiC*-merodiploid mutant MuD, and the IdiB-free mutant K10 104
  3.6.1 Growth of *S. elongatus* PCC 7942 WT, mutant K10 and mutant MuD under iron-sufficient and iron-depleted conditions 104
  3.6.2 Expression of the iron-regulated proteins IdiA, IdiB, and IdiC in *S. elongatus* PCC 7942 WT, mutant K10, and mutant MuD 105

3.7 Comparative transcript profiling of iron-dependent regulated genes in *S. elongatus* PCC 7942 WT, the *idiC*-merodiploid mutant MuD, and the IdiB-free mutant K10 106

3.7.1 Detection of transcripts of major iron-regulated clustered genes 108
3.7.2 Detection of transcripts encoding electron transport-related proteins (photosynthesis and respiration)..118
3.7.3 Detection of transcripts of genes encoding carbon metabolism-related proteins 119
3.7.4 Detection of transcripts of genes encoding nitrogen metabolism-related proteins 119
3.7.5 Detection of transcripts of genes encoding general stress proteins 119
3.7.6 Detection of transcripts of genes encoding regulatory proteins....................120
3.7.7 Major differences between *S. elongatus* PCC 7942 WT, mutant K10, and mutant MuD in the acclimation to iron limitation...121

3.8 Preliminary work: identification of putative transcription regulators controlling the transcritpion of the *idiB* operon in *S. elongatus* PCC 7942 ... 123

3.8.1 Identification of a novel Fur-homologous transcriptional regulator in *S. elongatus* PCC 7942 123
3.8.2 Identification of a novel MerR-like transcriptional regulator in *S. elongatus* PCC 7942 129

4 Discussion 133

4.1 Transcript-profiling of *S. elongatus* PCC 7942 and selected mutants grown under iron-deficient and iron-sufficient conditions...134
4.2 The functional role of IdiC in the modification of electron transport under iron starvation 139

5 Summary 142

6 Acknowledgements 144

7 References 145

8 Appendix 163

8.1 Supplementary tables for chapter 3.7..163
8.2 Abbreviations 166

1 Introduction

1.1 Cyanobacteria

1.1.1 General considerations about cyanobacteria

Cyanobacteria are a remarkably old monophylogenetic group of eubacteria, which have evolved probably almost 3.5 billion years ago (Schopf 2000). They were the first oxygenic photosynthetic organisms, and in evolutionary terms cyanobacteria represent the link between heterotrophically growing bacteria and algae as well as higher plants (Bradford 1976; Fay 1983). Cyanobacteria are characterized by their ability to synthesize chlorophyll *a* (Whitton and Potts 2000) and to perform an oxygenic type of photosynthesis (Carr and Whitton 1982). Their broad metabolic activity enabled the switch from an anoxygenic to present-day oxygenic atmosphere and allowed for the development of more complex eukaryotic life forms and more efficient catabolic pathways such as oxidative respiration.

Due to their considerable morphological diversity (Whitton and Potts 2000) and their great metabolic flexibility (Vermaas 2001), cyanobacteria are able to colonize virtually all terrestrial and aquatic habitats. Approximately 2000 cyanobacterial species have been described until now (Whitton and Potts 2000). In spite of their long evolutionary history, cyanobacteria still contribute up to 40% to the global biomass production and facilitate vast amounts of nitrogen fixation (Paerl 2000).

Today, cyanobacteria are an excellent research model system to investigate the principles of oxygenic photosynthesis. They grow relatively fast, are easy to handle, and can easily be genetically manipulated. In addition, several cyanobacterial genomes have been sequenced by now, allowing comparative bioinformatic and physiological analyses of different strains.

Synechococcus elongatus PCC 7942, formerly described as *Ancystis nidulans* and hereafter referred to as *S. elongatus* PCC 7942, represents an unicellular rod-shaped organism found in fresh water. Unlike many other cyanobacterial species, *S. elongatus* PCC 7942 is not able to fix molecular nitrogen. Its genome was entirely sequenced quite recently and is available at the JGI microbial genomics homepage (http://genome.ornl. gov/microbial/syn_PCC7942). The genome is composed of 2.7 mbp, has a GC-content of 55.4%, and contains 2653 annotated genes, which are supposed to encode for putative proteins. As a result of its photoautotrophically growth and the possibility of an easy genetical modification, *S. elongatus* PCC 7942 represents an excellent model system to study changes and modifications in photosynthetic electron transport to environmental changes and stresses.

1.1.2 Morphology of the cyanobacterial cell

Cyanobacteria are unicellular or multicellular, often filamentous prokaryotes and have no eukaryotic type of compartmentation. The prokaryotic cellular organization of cyanobacteria lacks membrane bound organelles but neither resembles closely the organization of Gram-negative bacteria nor the organization of Gram-positive bacteria (Gantt 1994).

However, the cyanobacterial cell wall structure strongly resembles those of Gram-negative eubacteria. The cell is surrounded by an external carbohydrate-enriched glycocalix being closely attached to the outer membrane. The glycocalix protects the cell against desiccation and represents a diffusion barrier for dissolved substances in the aqueous environment (Gantt 1994). In cyanobacteria, the cell wall consists of the outer membrane, a peptidoglycan layer, and the cytoplasmic membrane. The peptidoglycan layer has structural similarities to that of Gram-negative bacteria, but due to its diameter it has been classified independently (Jurgens and Weckesser 1985). Together with the thylakoid membrane, being partly attached to the cytoplasmic membrane, these membranes form the inner and outer periplasmic space, the cytoplasm, and the intrathylakoid space (see Figure 1.1).

In addition to the membrane systems, the cyanobacterial cell has a nucleosome in the centre of the cell which contains the DNA. Small gas-filled vesicles, called gas vacuoles, enable cyanobacteria to regulate their floating heights in water. These are especially important for obligate photoautotrophic cyanobacteria. Since cyanobacteria have no eukaryotic type of compartmentation, they have to convert a surplus of many metabolites into osmotically inactive forms leading to the formation of selected granules such as carboxysomes, polyphosphate granules, glycogen storage granules, and cyanophycin granules.

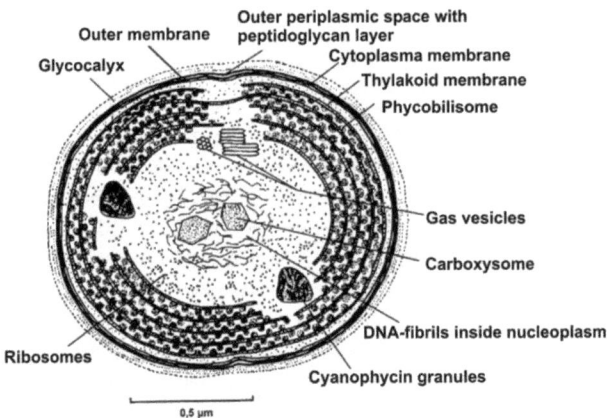

Figure 1.1: Schematic presentation of a cross section of a vegetative cyanobacterial cell (Van den Hoek 1995, modified).

Investigating the proteins located in the cytoplasmic membrane, the subunits of the ATP synthase, components of the respiratory electron transport chain, various transport systems, as well as partially assembled photosystem I (PS I), and photosystem II (PS II) complexes have been identified (Huang et al. 2002). Furthermore, proteins for assembly of the PS are located in the cytoplasmic membrane, while components of the proximal antenna of PS II, CP47 and CP43, are exclusively present in the thylakoid membrane (Zak et al. 2001). It has been proposed that the initial steps of PS assembly take place in the cytoplasmic membrane followed by a transfer of the partially-assembled complexes into the thylakoid membrane (Liberton et al. 2006).

1.2 Cyanobacterial electron transport

The electron transport system in cyanobacteria is highly sophisticated, because they contain a photosynthetic electron transport chain located in the thylakoid membrane and two distinct fully operational respiratory electron transport chains, which are located in the cytoplasmic membrane and in the thylakoid membrane (Schmetterer 1994). Thus, in the thylakoid membrane, both electron transport chains utilize the same redox components. This is the reason why the interrelationship between photosynthesis and respiration in cyanobacteria is especially complex and has to be strictly regulated to prevent futile cycles. The products of the light reactions, NADPH+H$^+$ and ATP, are subsequently used in the CO_2 fixation reactions by the Calvin cycle located in the cytoplasm (see Figure 1.2).

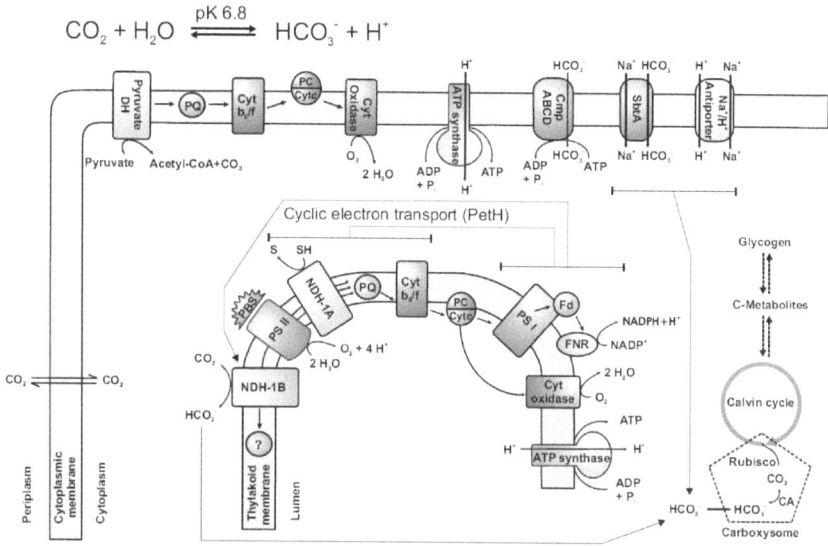

Figure 1.2: Model of the photosynthetic and respiratory electron transport chain in the thylakoid and in the cytoplasmic membrane of cyanobacteria as well as pathways of carbon and nitrogen metabolism ((Nodop et al. 2006); modified). CA = carboanhydrase, C-metabolites = carbon metabolites, Cyt b_6/f = cytochrome b_6/f complex, PC = plastocyanin, Cyt c = cytochrome c, Cyt oxidase = cytochrome oxidase, Cmp = bicarbonate transporter, DH = dehydrogenase, Fd = ferredoxin, FNR = ferredoxin:NADP oxidoreductase, NDH-1A = NADH dehydrogenase type A, NDH-1B = NADH dehydrogenase type 1 B, PS I = photosystem I, PS II = photosystem II, PQ = plastoquinone pool, SbtA = sodium-bicarbonate transporter.

Moreover, it should also be pointed out that the thylakoid membrane embedded electron transport chain is also extremely complex, because besides H_2O and NADH a number of additional electron donors can be utilized by the substrate dehydrogenases of the respiratory transport chain, such as H_2S, H_2, NADPH+H$^+$, succinate or L-amino acids. Additional electron acceptors besides NADP+ or O_2 are also possible, such as nitrate, nitrite, thioredoxin, H$^+$ and N_2.

1.2.1 Photosynthetic electron transport

Photosynthesis can be defined as the synthesis of organic compounds through fixation of carbon dioxide (CO_2) using light as an energy source. In cyanobacteria as oxygenic photosynthetic organisms, the light reactions of the photosynthetic process lead to the production of $NADPH+H^+$ and ATP, utilizing water as electron donor, while O_2 is released as a by-product (Ke 2001). The products $NADPH+H^+$ and ATP are subsequently used in the CO_2 fixation reactions of the Calvin cycle (Heinecke 2001; Tabita 1994) and at least in some cyanobacteria to a certain extent also in an alternative CO_2 fixation (via carbamoyl phosphate → citrulline → arginine; (Tabita 1994; Tabita 1987)). These light reactions also lead to the formation of an electrochemical proton gradient across the thylakoid membrane, which is the driving force for ATP synthesis by ATP-synthase (Frasch 1994; Lengeler et al. 1999). The light reactions of cyanobacterial oxygenic photosynthesis, mediating electron transport from water to $NADP^+$, are catalyzed by different membrane-bound protein complexes, being embedded in intracellular membrane systems (ICMs = intracellular membranes = thylakoid membranes) in most cyanobacterial species, and are connected by different soluble electron carriers (Vermaas 2001).

1.2.1.1 Photosynthetic linear electron transport

The photosynthetic electron transport chain of cyanobacteria contains two photosystems, PS II and PS I, which are coupled by a linear electron transport system consisting of the cytochrome b_6/f complex (Cyt b_6/f), and the mobile electron carriers plastoquinone (PQ) and cytochrome c_{553} (or plastocyanin (PC)).

PS II catalyses light-driven electron transfer from water to the PQ-pool. In each charge separation event, one electron is extracted from the Mn-containing oxygen evolving complex. If four positive charges have been accumulated, the electrons are refilled by oxidation of water to oxygen. The light excited electron is transferred from P_{680} to the Cyt b_6/f-complex by the mobile electron carrier PQ, which takes up two electrons and two protons and leaves PS II as PQH_2. This molecule diffuses through the membrane to the Cyt b_6/f-complex, which reduces the soluble electron carrier PC, located in the lumen of the thylakoids and in some cyanobacteria replaced by cytochrome c_{553}. The reduced PC donates the electron to PS I. PS I captures light energy to eject an electron from the reaction centre P_{700} using the excitations energy. The electron is transferred across the membrane to ferredoxin as a soluble electron carrier. The reaction is completed by reduction of the oxidized P_{700} by the reduced PC at the luminal side of PS I. Ferredoxin finally transfers the electron to the ferredoxin: $NADP^+$ oxidoreductase (FNR), which reduces $NADP^+$ to $NADPH+H^+$ (see Figure 1.3).

Linear photosynthetic electron flow:

H_2O ➡ PS II ➡ PQ pool ➡ Cyt b_6/f ➡ Cyt c_{553}/PC ➡ PS I ➡ Fd ➡ FNR ➡ $NADP^+$

Figure 1.3: Model of the photosynthetic linear electron transport in cyanobacteria with water as electron donor of photosynthesis. Abbreviations are the same as given in Figure 1.2.

The open chain electron transfer builds up an electrochemical proton gradient across the thylakoid membrane, which is used by ATP synthase for the synthesis of ATP from ADP and P_i (DeRuyter and Fromme 2008).

1.2.1.2 Photosynthetic cyclic electron transport

In addition to photosynthetic linear electron flow, the formation of reducing equivalents (NADPH+H$^+$) can be established without participation of PS II via a photosynthetic cyclic electron transport around PS I. This pathway does not lead to the proton dependent production of ATP via ATP synthase.

As shown in Figure 1.4, PS I cyclic electron transport has been proposed to consist of at least four different pathways (Jeanjean et al. 1999; Johnson 2005; Joliot and Joliot 2006; Yeremenko et al. 2005): (i) a NADH:plastoquinone oxidoreductase (NDH-1) dependent pathway first described by (Mi et al. 1994; Mi et al. 1992), (ii) a ferredoxin/flavodoxin-dependent ferredoxin:plastoquinone oxidoreductase (FQR) pathway with a participation of Ssr2016 as described for *Synechocystis* sp. PCC 6803 (Yeremenko et al. 2005), (iii) a pathway involving ferredoxin/flavodoxin-dependent ferredoxin:NADP$^+$ oxidoreductase (FNR) (Zhang et al. 2001), and (iiii) a pathway with a participation of ferredoxin as a direct substrate of the NDH-1 complex. It has been suggested that the most important role of PS I cyclic electron transport may be its contribution to stress tolerance to prevent irreversible photodamage. An increase in cyclic electron transport activity has been described for a number of stress conditions, such as e.g. high light (Herbert et al. 1995), salt stress (Joset et al. 1996), drought or low CO_2 (Deng et al. 2003; Golding and Johnson 2003).

Cyclic photosynthetic electron flow around photosystem I:

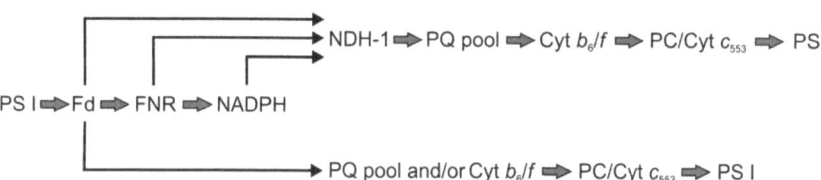

Figure 1.4: Model of the photosynthetic cyclic electron transport routes around PS I in cyanobacteria. Abbreviations are the same as in Figure 1.2.

However, the cyanobacterial NDH-1 complex plays a major role in PS I driven cyclic electron transport. Since the substrate binding subunits are so far not identified, the direct substrate is still unknown (see chapter 1.2.3). It has been suggested that NDH-1 becomes reduced directly by ferredoxin, displaced under several stress conditions by flavodoxin, or indirectly by ferredoxin/flavodoxin via FNR as well as by NADPH+H$^+$ (Mi et al. 1994; Mi et al. 1992) donating the electrons to the PQ-pool. From here the electrons were transferred back to PS I according to linear electron transport via the Cyt b_6/f-complex and PC.

Although photosynthetic cyclic electron transport has been investigated extensively for years, it still in part remains enigmatic, since not all components have been identified and thus, the number of possible pathways is still unknown. The identification of the substrate binding subunits of the cyanobacterial NDH-1 complex plays a major role for detecting the direct substrate.

1.2.1.3 Photosystem I

As mentioned above, plastocyanin and/or cytochrome c_{553} are soluble electron carriers transporting electrons from the Cyt b_6/f complex to the oxidized PS I, which can be defined as a light-driven plastocyanin/cytochrome c_{553} ferredoxin oxidoreductase. The primary chlorophyll donor or reaction centre chlorophyll of PS I is a special pair chlorophyll called P_{700}, which becomes excited directly by light or excitation energy from the antenna systems containing 50 to 100 chlorophyll molecules.

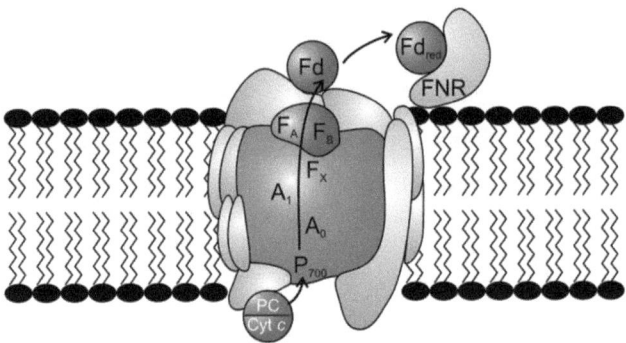

Figure 1.5: Simplified presentation of the cyanobacterial PS I complex. Abbreviations as given before, P_{700} = special pair reaction centre chlorophyll a, A_0 = chlorophyll monomer, A_1 = phylloquinone, F_X = [4Fe-4S]-cluster, F_A = [4Fe-4S]-cluster, F_B = [4Fe-4S]-cluster (Michel, unpublished).

After illumination, an electron is transferred from an excited state of P_{700} to the stromal side of PS I via the primary acceptor A_0 (a chlorophyll monomer), the secondary acceptor A_1 (a phylloquinone) and three [4Fe-4S]-clusters F_X, F_A, and F_B (see Figure 1.5). In contrast to higher plants, PS I of cyanobacteria has been reported to exist in a monomeric and a trimeric form. The molecular mass of PS I trimers, containing about 110 chlorophyll a molecules, is about 900 kDa. A trimeric structure of PS I favours cyclic electron transport around P_{700}, whereas monomeric PS I is more active in linear electron transport (Meunier et al. 1997). A shift from monomeric to trimeric PS I can be traced via 77K chlorophyll a fluorescence, since monomeric PS I shows a 1.75-fold higher fluorescence compared to trimeric PS I at 715 nm (Bruce and Biggins 1985; Bruce et al. 1985; Golitsyn et al. 1995; Meunier et al. 1997).

The three-dimensional structure of PS I trimers from the thermophilic cyanobacterium *Thermosynechococcus elongatus* BP-1 has recently been published at 2.5 Å resolution. This crystal structure provides a picture at atomic detail of 11 protein subunits and 127 cofactors comprising 96 chlorophylls, 2 phylloquinones, 3 [4Fe-4S] centres, 22 carotenoids, 4 lipids, a putative Ca^{2+} ion, and 201 water molecules (Fromme and Witt 1998; Jordan et al. 2001). The cyanobacterial subunits of PS I denoted PsaA to PsaF and PsaI to PsaM are given in Table 1.1.

Table 1.1: Protein subunits of cyanobacterial PS I (Golbeck 1994; Krauß and Sänger 2001). "TM" transmembrane helices.

Genes	Proteins	M.M. [kDa]	TM-Helices	Comments
psaA	PsaA	83	11	Binds 110 Chl a molecules, antenna chlorophyll and photochemical charge separation
psaB	PsaB	83	11	Binds ß-carotene molecules, 2 phylloquinones (Vit K1), and F_x [4Fe-4S], involved in antenna and photoprotection
psaC	PsaC	9	0	Located on the stromal side, F_A and F_B [4Fe-4S]
psaD	PsaD	15	0	Located on the stromal side, ferredoxin docking and PsaC binding
psaE	PsaE	8	0	Located on the stromal side, involved in ferredoxin/flavodoxin docking, cyclic electron flow
psaF	PsaF	15	1-2	Plastocyanin/Cyt c_6 docking
psaI	PsaI	4	1	Stabilizes PsaL in PS I complex
psaJ	PsaJ	5	1	Stabilizes PsaL in PS I complex
psaK	PsaK	8	2	Function unknown
psaL	PsaL	16	2-3	Function in PS I trimerization
psaM	PsaM	3	1	Function unknown

1.2.1.4 Ferredoxin:NADP$^+$ oxidoreductase

In cyanobacteria and chloroplasts the main function of this FAD-containing enzyme is to catalyse the final step of linear photosynthetic transport, catalysing the electron transfer from reduced ferredoxin to NADP$^+$, and thus providing NADPH+H$^+$ for CO_2 assimilation and other reductive metabolism. In non-photosynthetic root plastids a genetically distinct FNR isoform is postulated to function in the opposite direction, providing electrons for nitrogen assimilation at the expense of NADPH+H$^+$ generated by heterotrophic metabolism (Neuhaus and Emes 2000). Besides its function in reducing NADP$^+$ using electrons from reduced ferredoxin, FNR has been reported to oxidize with concomitant reduction of several different substrates such as ferricyanide, indophenol (diaphorase activity), Cyt f, Cyt c, plastocyanin, and quinols (see review: (Schmetterer 1994)). These activities strongly support the idea that FNR may act in addition as a NADPH+H$^+$ dehydrogenase in the respiratory electron transport pathway of oxygenic photosynthetic organisms (Guedeney et al. 1996; Scherer et al. 1988).

Despite the similarities in both biochemical properties and protein structure for the cyanobacterial and the plastid FNR (Karplus and Faber 2004), the unique *petH* gene encodes in most phycobilisome (PBS)-containing cyanobacteria a 46 kDa FNR that contains an N-terminal CpcD linker domain of ~ 80 aa, whose sequence is similar to PBS rod-linker polypeptides (Nakamura et al. 2003; Schluchter and Bryant 1992; Steunou et al. 2006). The PBS are large and abundant bilin-protein complexes which harvest light for photosynthesis. During stress conditions, e.g. iron or nitrogen starvation, PBS degradation supplies amino acids for the synthesis of essential proteins (Allen and Smith 1969). Purified PBS from several cyanobacterial strains contain significant amounts of the 46 kDa FNR, in particular one to two molecules per PBS (Schluchter and Bryant 1992; van Thor et al. 1999; Yamanaka et al. 1978). It was proposed that the FNR binds to the PBS rods via its linker domain to fulfil functions in both cyclic electron transport and respiration, in close proximity to the thylakoid membrane (Gomez-Lojero et al. 2003; van Thor et al. 2000). In contrast, a 35 kDa FNR without the N-terminal linker domain has been purified from several cyanobacterial strains (*Spirulina sp.* (Yao et al. 1984), *Anabaena cylindrical* (Rowell et al. 1981), and *Synechocystis* sp. PCC 6803 (Matsuo et al. 1998)). It is suggested that this two domain FNR retains full functionality in its classical function in the last step of linear photosynthetic electron transport transferring electrons from ferredoxin to $NADP^+$ (Matthijs et al. 2002).

It is suggested, that the soluble two domain FNR of *Synechocystis* sp. 6803 is a result of proteolytic degradation of the N-terminal domain of 46 kDa FNR, using a PEST cleavage site (Matthijs et al. 2002). In contrast, it has been shown recently that the large FNR isolated from *Synechocystis* sp. 6803 undergoes proteolysis when not bound to PBS, whereas the functional 35 kDa isoform is produced from an internal ribosome entry site within the *petH* ORF. This isoform accumulates under conditions of heterotrophic metabolism and occurs therefore only in cyanobacteria capable of heterotrophic growth (Thomas et al. 2006).

1.2.2 Respiratory electron transport chain

Respiration has been defined as a bioenergetic process, which leads to the production of reducing equivalents ($NADPH+H^+$) and energy equivalents (ATP) by the oxidation of organic carbon compounds (Schmetterer 1994). The cyanobacterial respiratory electron transport chain, located in the cytoplasmic and the thylakoid membrane, consists of several electron donating complexes such as the NADH:plastoquinone oxidoreductase (NDH-1), the succinate dehydrogenase (S-DH), and other different substrate dehydrogenases. These complexes are connected with the terminal cytochrome oxidases by the Cyt b_6/f complex using the mobile electron carriers plastoquinone, cytochrome c_{553} or plastocyanin, and cytochrome $_M$ (see Figure 1.6). The terminal electron transfer to oxygen is catalyzed by the cytochrome oxidases c or bd (Vermaas 2001).

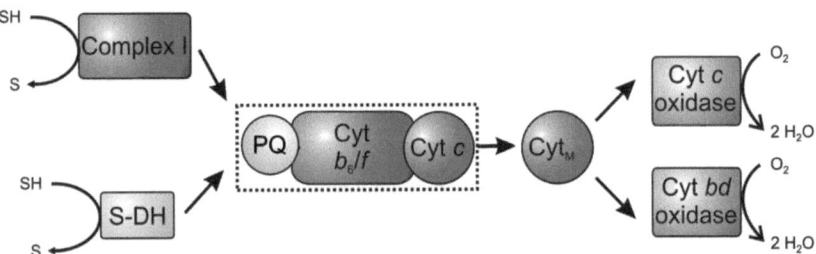

Figure 1.6: Model of the respiratory electron transport in the thylakoid membrane (NDH-1/substrate-DH → O_2). Complex I = NDH-1 complex, S-DH = substrate dehydrogenase, e.g. the succinate dehydrogenase, PQ = plastoquinone pool, Cyt b_6/f = cytochrome b_6/f complex, Cyt c = cytochrome c, Cyt $_M$ = cytochrome M, Cyt c oxidase = cytochrome c oxidase, Cyt bd oxidase = cytochrome bd oxidase (Michel 2003). In addition to Cyt c_{553}, Cyt $_M$ may participate in respiratory electron transport to terminal oxidases (Manna and Vermaas 1997).

The respiratory electron transport chain generates a proton gradient across the cytoplasmic membrane actuating the ATP-Synthase. The produced ATP is used by components of several essential metabolic pathways, e.g. CO_2 fixation or glycolysis.

Many open questions exist with respect to the cyanobacterial respiratory electron transport chain, since respiratory and photosynthetic electron transport chains are not physically separated and due to the complexity of the respiratory chain (Hart et al. 2005; Schmetterer 1994; Vermaas 2001). It has been shown that cyanobacteria depend on respiration in the dark to maintain energy levels and that respiration increases under stress conditions, such as high salt growth conditions (Fry et al. 1986) or iron deficiency (Michel 2003). Another level of complexity is added when it is taken into consideration that NDH-1 complexes exist in multiple forms (see chapter 1.2.3) (Battchikova et al. 2005; Herranen et al. 2004).

1.2.3 The cyanobacterial NDH-1 complex

Mainly related to the NDH-1 complex in cyanobacteria, there exist a number of open questions, which should be briefly addressed below, since some of these aspects are relevant for the interpretation of results presented in this thesis. The NDH-1 complex corresponds to the complex I of the respiratory electron transport chain and is well characterised in several bacteria, such as *E. coli*, and in mitochonderia of eukaryotes (Friedrich 1998; Grigorieff 1999; Weidner et al. 1993). The enzyme complex is composed of at least 14 subunits (like in *E. coli*: NuoA to NuoN) and as many as 40 subunits in bovine mitochondria. The complex I consists of a membrane module interacting with the quinone, a further membrane module interacting with the plastoquinone binding module, and a substrate binding module. The latter binds and oxidises NADPH+H$^+$. It contains the prosthetic group FMN and several Fe-S-centres (Friedrich 1998; Friedrich et al. 1995). Possibly NDH-1 initially was a ferredoxin:quinone oxidoreductase and eventually in evolution emerged by acquisition of the NADH dehydrogenase module to become a NADH:quinone oxidoreductase (Berger et al. 1993; Friedrich and Scheide 2000).

In *Synechocystis* sp. PCC 6803 12 *ndh* genes with similarity to the genes *nuoA-D* and *nuoH-N* of *E. coli* have been identified (Berger et al. 1993; Friedrich and Scheide 2000; Kaneko et al. 1996). The *ndh* genes of *Synechocystis* sp. PCC 6803 and their homologous in *E. coli* are given in Table 1.2.

Table 1.2: Nomenclature of the subunits of the cyanobacterial complex I and complex I from *E. coli* (Friedrich & Scheide 2000). [a] The genes of NuoC and NuoD are fused to one gene NuoCD in *E. coli* (Friedrich et al. 1997), and [b] NuoC of *E. coli* contains an additional [2Fe-2S] cluster (Friedrich et al. 1995).

Gene	Protein	E. coli homologue	Localization	Predicted function
ndhC	NdhC	NuoA	membraneous	-
ndhK	NdhK	NuoB	peripheral	[4Fe-4S]-cluster
ndhJ	NdhJ	NuoC[ab]	peripheral	-
ndhH	NdhH	NuoD[a]	peripheral	Proton channel; [NiFe] cluster
ndhA	NdhA	NuoH	membraneous	Quinone binding
ndhI	NdhI	NuoI	peripheral	2x [4Fe-4S]-cluster
ndhG	NdhG	NuoJ	membraneous	-
ndhE	NdhE	NuoK	membraneous	-
ndhF	NdhF	NuoL	membraneous	Proton translocation
ndhD	NdhD	NuoM	membraneous	Proton translocation
ndhB	NdhB	NuoN	membraneous	Proton translocation
ndhL	NdhL	-	-	Proton translocation

The membrane-bound module of the NDH-1 complex is build-on seven protein subunits, while four proteins constitute the connecting module in between the membrane-bound module and the substrate-binding module (see Figure 1.7). However, genes homologous to the substrate binding-subunits NuoE and NuoF have not been identified in cyanobacteria so far (Kaneko et al. 1996; Kösling 1999). Thus, the substrate used by the cyanobacterial NDH-1 complex remains unknown (Schmetterer 1994).

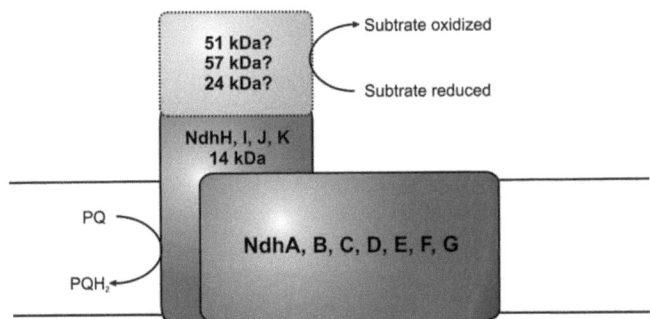

Figure 1.7: Schematic representation of the cyanobacterial NDH-1 complex (Berger et al. 1993).

In general the *ndh* genes in cyanobacteria are present in one copy, while five homologous genes for *ndhD* and four homologous genes for *ndhF* are present (Kaneko et al. 1996). The five NdhD proteins differ in their amino acid sequence with highest similarity between NdhD1 and NdhD2 as well as NdhD3, NdhD4, and NdhD5. Therefore, different types of NDH-1 complexes have been identified. These complexes vary in NdhD and NdhF subunits leading to different suggested functions (Herranen et al. 2004; Ohkawa et al. 2000a; Ohkawa et al. 2000b).

Previous studies referred to a NDH-1A complex consisting of NdhF1, NdhD1 or NdhD2 in addition to the single copy proteins. Its assigned function is the transfer of electrons from NADPH+H$^+$ or another substrate to plastocyanin in the respiratory electron transport chain as well as in cyclic electron flow around PS I (see Figure 1.8) (Klughammer et al. 1999; Mi 1992; Mi 1994; Mi 1995; Ogawa and Kaplan 1987; Ohkawa et al. 2000a; Ohkawa et al. 2000b; Schmetterer 1994). In addition, evidence was found for a NDH-1B complex consisting of NdhF3 or NdhF4 and NdhD3 or NdhD4 next to the single copy proteins. This complex was supposed to function in inducible and constitutive CO_2 uptake as well as in the conversion of CO_2 to HCO_3^- (Badger and Price 2003; Klughammer et al. 1999; Ogawa 1991b; Ogawa 1991a; Ohkawa et al. 2000a; Ohkawa et al. 2000b; Ohkawa et al. 2001; Shibata et al. 2001).

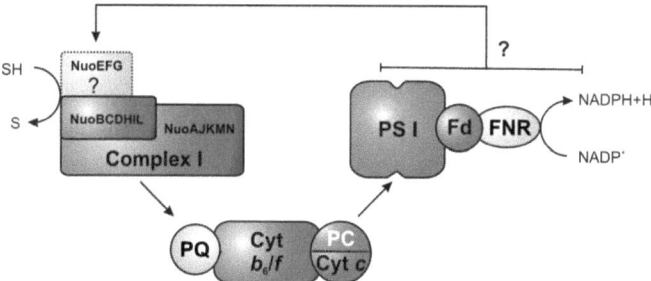

Figure 1.8: Schematic presentation of cyanobacterial complex I and its proposed function in cyclic photosynthetic electron transport around photosystem I. Abbreviations as given in previous figures and tables. NuoEFG have not yet been identified in cyanobacteria. Thus, other and multiple substrate-binding modules might exist, which are expressed to allow adaptation to different environmental conditions and metabolic activities (Michel, unplubished).

More recent studies have identified three different complexes, the NDH-1L, the NDH-1M, and the NDH-1S complex. NDH-1L and NDH-1M consist of 10 subunits (NdhA, B, C, G, H, I, J, K, Sll1262, and Slr1623) (Prommeenate et al. 2004; Zhang et al. 2004). NDH-1L contains NdhD1/D2 and NdhF1 in addition. NDH-1S is a small complex with a simpler protein composition comprising NdhD3, NdhF3, CupA and Sll1735 (Zhang et al. 2004). NDH-1L has been assigned to support heterotrophic growth and to mediate in respiratory electron transport, while NDH-1M participates in cyclic electron flow activity around PS I and inducible CO_2 uptake. NDH-1S plays a role in inducible CO_2 uptake, too (Battchikova 2008; Zhang et al. 2004).

Apart from the type 1 NDH, cyanobacteria possess a second type of NADH dehydrogenase. The so called type 2 dehydrogenase (NDH-2) consists of a single polypeptide chain (Howitt et al. 1999). The characteristics of the eubacterial enzyme have been reviewed extensively (Yagi 1993; Yagi 1991; Yagi 1998). *Synechocystis* sp. PCC 6803 possesses three genes for NDH-2 (*ndbA*, *ndbB*, and *ndbC*; (Kaneko et al. 1996)). NDH-2 has never been shown to contribute significantly to respiratory activity. Thus, it is suggested to have a main role as redox sensor, e.g. detecting the redox state of the PQ-pool (Howitt et al. 1999).

In contrast to the thylakoid membrane, the respiratory electron transport chain in the cytoplasmic membrane lacks any type of NDH-1 complex. Thus, the major substrates providing electrons for this pathway are still uncertain. Possible candidates are succinate via the succinate dehydrogenase (Vermaas 2001), pyruvate via the pyruvate dehydrogenase (Engels et al. 1997), L-amino acids via L-amino acid dehydrogenases (Anraku and Gennis 1987; Bockholt et al. 1996), and NAD(P)H via NDH-2s.

1.2.4 L-amino acid oxidase

S. elongatus PCC 7942 possess the gene *aoxA* coding for an L-amino acid oxidase (L-Aox) (Bockholt et al. 1995). A similar L-Aox was detected in *Synechococcus cedrorum* PCC 6908 (Gau et al. 2007), and evaluation of the genome of *Synechocystis* sp. PCC 6803 revealed the presence of a gene which has similarity to the *aoxA* gene of *S. elongatus* PCC 7942. The L-Aox has an unusual substrate specificity. It catalyses the oxidative deamination of basic L -amino acids (L-Arg>L-Lys>L-Orn>L-His). Another unusual property besides having this rather unusual substrate specificity for basic L-amino acids, is its strong inhibition by cations ($M^{3+} > M^{2+} > M^+$) and less strong inhibition by anions. In the group of divalent cations, transient metal ions like Mn^{2+} inhibit more strongly than alkali earth metal ions like Ca^{2+} (Engels et al. 1992; Pistorius et al. 1979; Pistorius and Voss 1980). The L-Aox is mainly located in the soluble protein fraction of the periplasm, however, in part the L-Aox is associated with the thylakoid membranes (Bockholt et al. 1996). Therefore, one possible role might be that the enzyme has a function in mediating electron flow from L-arginine oxidation to the respiratory/photosynthetic electron transport chain and thus represents an alternative substrate dehydrogenase for the electron transport chain (see Figure 1.9). Present work of our group supports such an assumption (Schriek 2008).

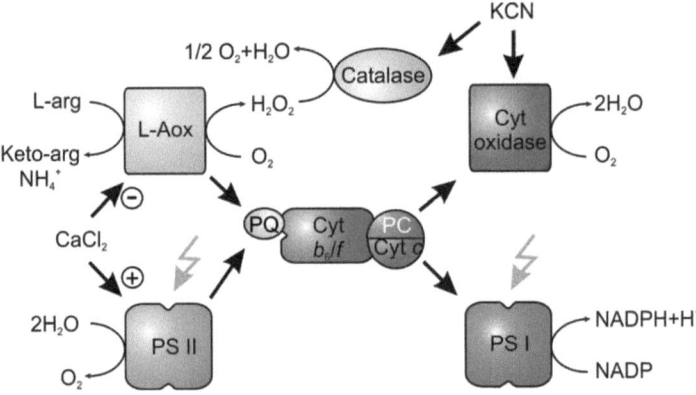

Figure 1.9: L-arginine as a possible alternative substrate of the electron transport chain: The antagonistic effect of $CaCl_2$, L-arginine oxidation (inhibitory) and water oxidation (stimulatory). Abbreviations as given in previous figures and tables; L-arg = L-arginine.

This model of the L-Aox being an alternative substrate dehydrogenase, which transfers electrons from the oxidation of L-arginine into the electron transport chain, is quite interesting because of the special role that L-arginine has in the cyanobacterial metabolism: synthesis of cyanophycin (multi-L-arginyl-poly-aspartic acid) (Simon 1987) and partial CO_2 fixation via carbamoyl phosphate leading to citrulline and subsequently arginine (Tabita 1994).

With thylakoid membrane fractions of *S. elongatus* PCC 6301/PCC 7942, it could be demonstrated that $CaCl_2$ had an antagonistic effect on photosynthetic water oxidation (stimulatory) and L-arginine oxidation by the L-Aox (inhibitory) suggesting that water and L-arginine might be alternative substrates for the electron transport chain in the thylakoid membrane.

1.2.5 Interrelationship between photosynthesis and respiration

As described above, linear photosynthetic and respiratory activities are combined in a single compartment in cyanobacteria. In addition, cyclic electron transport around PS I takes place in the same membrane and in part uses the same redox carriers and redox-active components. This suggests a strong relationship between photosynthesis and respiration (see Figure 1.10).

Introduction

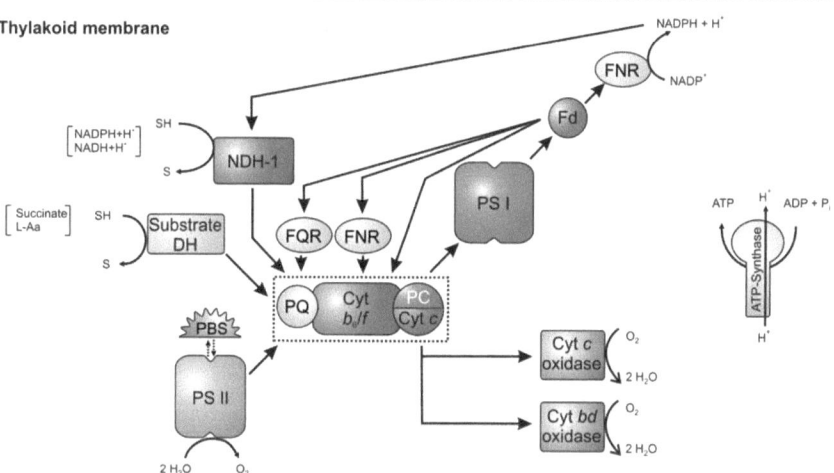

Figure 1.10: Model of the electron transport chain in the cytoplasmic and thylakoid membrane of cyanobacteria. Photosynthetic electron transport chain (thick arrows): PS II, cytochrome b_6/f complex (Cyt b_6/f), plastoquinone (PQ), cytochrome c_{553}/plastocyanin (Cyt c/PC), ferredoxin (Fd), ferredoxin:NADP oxidoreductase (FNR), and PS I. Respiratory electron transport chain (thin arrows): NAD(P)H-dehydrogenase (NDH), succinate dehydrogenase (Succ-DH), plastoquinone (PQ), cytochrome b_6/f complex (Cyt b_6/f), cytochrome c_{553}/plastocyanin (Cyt c/PC), cytochrome oxidase (Cyt-Oxidase). Additional donors: H_2S (sulfide:quinone oxidoreductase; SQR), and H_2 („uptake" hydrogenase; HUP). Additional acceptors: NO_3^-, NO_2^-, thioredoxin, H^+, N_2, and O_2 by alternative quinol oxidase (QOX) (Michel 2003).

A major complexity may arise from the high numbers of substrate dehydrogenases, which have the potential to feed electrons to the PQ pool. Beside the NDH-1 complex the succinate dehydrogenase is able to oxidise succinate transferring the electrons to the PQ pool. The contribution of other substrate dehydrogenases to plastoquinone pool reduction, however, is still discussed controversially (Ohkawa et al. 2000). Another substrate dehydrogenase, the pyruvate dehydrogenase, which catalyses the oxidation of pyruvate and subsequent reduction of $NADPH+H^+$, may also donate electrons to the PQ pool. An interaction of this enzyme with the respiratory electron transport chain (Engels et al. 1997; Engels and Pistorius 1997) and NDH-1 has been postulated (Sumegi and Srere 1984). Amino acid dehydrogenases may also represent alternative electron input devices. The cyanobacterial L-Aox has been suggested to mediate the electron flow from L-Arginin to the PQ pool (Bockholt et al. 1995; Bockholt et al. 1996; Engels et al. 1992).

The cyanobacterial NDH-1 is involved in respiration and possibly in cyclic electron transport around PS I. FNR and Cyt b_6/f complex are shown to be involved in linear as well as in cyclic electron transport. Moreover, ferredoxin is the electron donor for nitrogen assimilation reactions and the thioredoxin system, which mediates light-control of metabolic functions.

Therefore, all activities must be strictly regulated in dependency of the environmental conditions to adjust the cellular metabolic status, since they are in part counteracting and would result in futile cycles when taking place at the same time. Although it is widely accepted that the redox status of the plastoquinone pool as well as the redox status of the Cyt b_6/f complex are important in regulation (Allen 1995; Allen et al. 1995; Allen and Nilsson 1997), the knowledge on regulatory means mediating redox control from the PQ pool and the ferredoxin-gating mechanism, distributing electrons from ferredoxin to various substrates, is scarce.

1.3 Adaption to iron limitation in cyanobacteria

Cyanobacteria have developed a large number of effective mechanisms to recognize and sense variations in their environment and to adapt to changes in light quantity, light quality, availability of macro- and micronutrients, temperature, and osmotic conditions (Anderson et al. 1995; Bhaya et al. 2000; Grossman et al. 1994; Mann 2000). In this chapter, the different adaptational strategies of cyanobacteria to iron limitation will be discussed.

1.3.1 The biological use of iron

Virtually almost all organisms except lactobacilli (Archibald 1983) have an absolute requirement for iron, and it has been suggested that reactions catalysed by iron may have constituted a first step in the origin of life (Cody et al. 2000; Wächtershäuser 1990; Wächtershäuser 2000). Approximately 5% of the earth's crust is composed of iron, and this great abundance coupled with the fact that the metal has two readily convertible redox states, has led to its evolutionary selection for an astonishing array of diverse biological reactions spanning a range from +300 up to -500 mV. Thus, the essential role of iron in cell growth and replication is due to the fact that many iron-containing proteins catalyse key reactions involved in photosynthesis, respiration, nitrogen assimilation, and DNA synthesis (Boyer et al. 1987; Straus 1994). Iron exerts its function either in heme or iron-sulfur proteins, or in proteins in which iron is exclusively bound via amino acid residues (Frausto da Silva and Williams 1993). However, there are two major problems related to the use of iron as an important element for essential biological reactions:

First, although iron is the fourth most abundant element by weight in the earth's crust, the bioavailability of iron is severely restricted. This is due to the fact that in aqueous oxygen containing environments Fe^{2+} is quickly oxidized to Fe^{3+}. At physiological pH, Fe^{3+} forms insoluble or only poorly soluble hydroxides and hydroxy-aquo complexes (Ratledge and Dover 2000), which reduce the biological availability of iron. Extended investigations showed that iron deficiency quite frequently occurs in marine habitats and often is the only nutrient limitation in otherwise nutrient-rich habitats. Over the past decade, a great deal of attention has been spent on the investigation of iron hypothesis, which states that low iron availability limits biomass production in vast regions of the world's oceans (Behrenfeld and Kolber 1999; Geider and La Roche 1994; Martin 1992; Martin et al. 1994; Martin and Fitzwater 1988).

Thus, the idea came up that a fertilization with iron (Martin et al. 1994) could significantly enhance biomass production and be a putative relief in terms of the global green house dilemma (Tortell et al. 1999).

Second, a close relationship between iron homeostasis and oxidative stress exists. In phototrophic as well as in heterotrophic organisms the lack of iron leads to the incomplete reduction of molecular oxygen and therefore, to formation of reactive oxygen species (ROS) (Mongkolsuk and Helmann 2002). The accumulation of ROS generates oxidative stress affecting function of enzymes and cell components. On the other hand, free intracellular iron also causes problems generating highly active hydroxyl radicals from hydrogen peroxide in the Fenton-Haber-Weiss reaction (see chapter 1.3.2) (Elstner 1990).

Thus, microorganisms had to develop highly effective systems for iron assimilation from an environment in which iron is poorly present in a soluble form as well as for the intracellular iron storage and effective further processing.

1.3.2 Interrelationship between iron homeostasis and oxidative stress

Especially for photoautotrophic organisms, the availability of iron is vitally important. In particular, the photosynthetic electron transport chain and also the respiratory chain contain a high number of iron-containing proteins. Thus, iron starvation has a large impact on the function of the electron transport chains.

Although iron is not a component of the light-harvesting phycobilisome complex, it plays an important role in the synthesis of phycobilin chromophores, because it is the cofactor of the heme oxidase. PS II contains a non-heme iron between Q_A, the primary electron accepting PQ that is bound to D2, and Q_B, the PQ bound to D1. In addition, PS II contains two Cyt b_{559} located on the subunits PsbE/PsbF and PsbV contains one Cyt c_{550}. Electrons, transferred to PQ from PS II, are subsequently passed to the Cyt b_6/f complex. This complex consists of four major subunits: Cyt b_6 which contains two hemes, cytochrome f which is a c-type cytochrome with one heme, a Rieske iron-sulfur protein that has a [2Fe-2S] centre, and a 17 kD peptide with no cofactor. PS I contains three [4Fe-4S] centres (F_X, F_A and F_B) that participate in electron transfer within the complex to a loosely attached ferredoxin molecule containing one [2Fe-2S] centre. Cyanobacteria which use cytochrome c_{553} instead of plastocyanin contain one additional heme. Thus, a minimum of 23 to 24 iron atoms is needed in the photosynthetic electron transport chain as shown in Figure 1.11 (Ferreira and Straus 1994; Straus 1994).

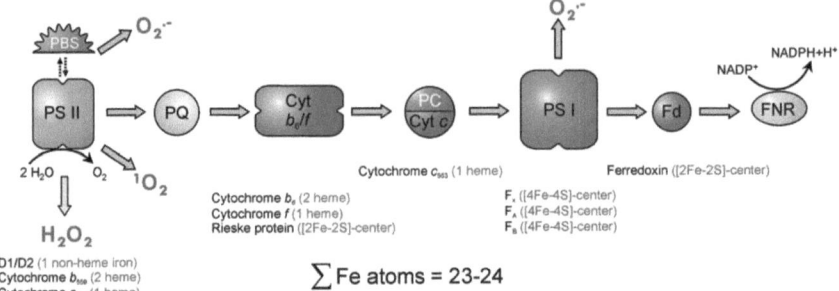

Figure 1.11: Iron cofactors involved in photosynthetic linear electron transport from H_2O to $NADP^+$ (Michel and Pistorius 2004). Abbreviations are given in the legend of Fig. 1.2 and in the text.

Once impaired by the lack of iron, the damaged photosynthetic electron transport chain reactions lead to a rapid and increased generation of ROS (Aro et al. 1993; Asada 1994; Mittler 2002). Mainly PS II and PS I contribute to the production of ROS significantly (Aro et al. 1993; Asada 1994; Asada 1999; Asada 2000). Superoxide anions can be produced via the interaction of O_2 with the acceptor side of PS I and PS II. Moreover, singulet dioxygen ($^1O_2$) production can result from the interaction of O_2 with excited triplet-state chlorophyll (three Chl a) in PS II and PS I (Krieger-Liszkay and Rutherford 1998; Rutherford and Krieger-Liszkay 2001; Trebst et al. 2002). H_2O_2 can be formed due to incomplete water oxidation on the donor side of PS II (Samuilov 1997; Wydrzynski et al. 1989) as well as by dismutation of superoxide anion by superoxide dismutase (Herbert et al. 1992).

Since several ROS-detoxifying enzymes, such as catalase, peroxidase or some superoxide dismutases require iron as cofactor, it is inevitable that iron limitation leads to a decreased decomposition of ROS (Lundrigen et al. 1997) and thus, results in an even more severe oxidative stress. This stress causes damage to the photosystems, especially PS II (Aro et al. 1993; Bhaya et al. 2000). However, not only a limitation of iron is a problem, free intracellular iron can also cause the formation of highly reactive hydroxyl radicals from H_2O_2 (Elstner 1990; Guerinot and Yi 1994). The Haber-Weiss reaction ($^{\cdot}O_2 + H_2O_2 \rightarrow O_2 + OH^- + HO^{\cdot}$) is the sum of the reduction of Fe^{3+} by superoxide anions ($^{\cdot}O_2 + Fe^{3+} \rightarrow O_2 + Fe^{2+}$) and the Fenton reaction ($H_2O_2 + Fe^{2+} \rightarrow Fe^{3+} + OH^- + HO^{\cdot}$).

Moreover, superoxide attacks iron-sulfur clusters and releases free iron, which further increases oxidative stress. Therefore, it is quite obvious that iron homeostasis and oxidative stress as well as redox regulation are closely intertwined as already described for eubacteria (Helmann 1998). This makes any consideration on adaptational mechanisms to iron limitation very complex, especially in photosynthetic organisms with minor compartmentation, such as cyanobacteria. The knowledge about adaptation to oxidative stress is rather limited in cyanobacteria (Mittler and Tel-Or 1991; Mittler and Tel-Or 1991a; Tichy and Vermaas 1999).

1.3.3 Consequences of iron limitation in cyanobacteria

Since primordial cyanobacteria were probably the first organisms that carried out an oxygenic type of photosynthesis (Schopf 2000; Whitton and Potts 2000), they were also the first organisms to be confronted with the interrelationship of iron limitation and oxidative stress. To cope with iron limitation, strategies have been developed that were classified into three categories by Straus (Straus 1994): (a) Acquisition, describing the excretion of siderophores and optimization of cellular iron transport systems, (b) Compensation, describing the replacement of dispensable iron-containing proteins or proteins requiring iron for their synthesis by other proteins without an iron cofactor, and (c) Retrenchment, describing the simple reduction of strictly iron-dependent metabolic activities, e.g. the reduction of pigments and thylakoid membranes. In addition to these three adaptational mechanisms the expression of iron-regulated proteins having a protective function has recently obtained considerable attention (see chapter 1.3.5).

Prolonged iron starvation as any other nutrient limitation will lead to severe alterations in cellular structures and to a reduction of physiological activities. In cyanobacteria the most obvious alterations under iron starvation are a reduction of the phycobilisomes and chlorophyll-containing proteins as shown for *S. elongatus* PCC 7942 and *Synechocystis sp.* PCC 6803 (see e.g.: (Guikema and Sherman 1984; Odom et al. 1993; Öquist 1971; Öquist 1974; Riethman and Sherman 1988; Sherman and Sherman 1983); (Straus 1994)). In addition to the reduction of pigment content a three- to four-fold reduction in the number of thylakoid membranes is seen in electron micrographs of cross sections of *S. elongatus* PCC 7942 cells (Sherman and Sherman 1983). Cells cultivated under iron-sufficient and iron-deficient conditions, are compared in Figure 1.12.

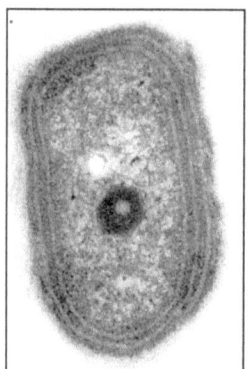
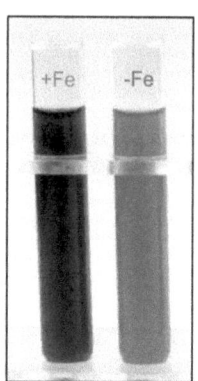
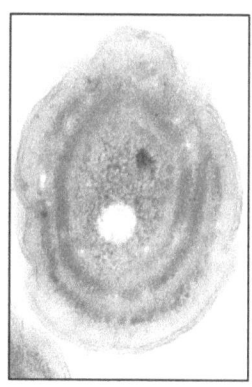

Iron-sufficient growth (2 days)　　　　　　　　　　　　Iron-deficient growth (2 days)

Figure 1.12: Consequences of iron limitation on the morphology of *S. elongatus* PCC 7942. Electron micrographs of iron-sufficient and iron-starved cells. Iron starvation leads to reduction of thylakoids and a loss of pigments including chlorophyll a and phycobilins within less than 48 hours (Pietsch 2004).

The reduction of pigments and thylakoids is paralleled by changes in the spectral properties of the chlorophyll proteins, which are directly correlated with a partial loss of high molecular mass chlorophyll protein complexes and the induction of a chlorophyll protein complex, called CPVI-4 (Riethman and Sherman 1988).

In addition to the reduction in pigment content, all iron-containing components of the photosynthetic electron transport chain have been reported to be reduced under iron depletion in several cyanobacteria (Ivanov et al. 2000; Sandmann 1985).

1.3.4 Iron acquisition and storage in cyanobacteria

The limited bioavailability of iron in an oxygenic environment has forced microorganisms to evolve two main strategies for successful scavenge of iron from their habitats for survival. One involves synthesis and secretion of low molecular mass, iron-binding chelators known as siderophores (sider, *gr.* = iron, phore, *gr.* = bearer) as well as transport of the chelated iron. Siderophores are structurally diverse (at least five different classes have been defined), but are often either catechol- or hydroxamate-containing compounds (Drechsel and Winkelmann 1997; Neilands 1995). The other system involves reduction of ferric to ferrous iron by a plasma membrane redox system, followed by uptake using specific transporters (Guerinot and Yi 1994; Von Wiren et al. 1994).

E. coli has specific receptor proteins in the outer membrane that bind ferrichrome (FhuA), ferric aerobactin (IutA), ferric coprogen or ferric rhodotorulate (FhuE), and ferric dicitrate (FecA). FhuA, FhuE, and IutA are components of siderophore-mediated iron transport systems that involve typical ATP binding cassette (ABC)-type transporters consisting of a periplasmic iron-binding protein (FhuD) and cytoplasmic membrane proteins (FhuB and FhuC) (Braun et al. 1983). Ferric dicitrate is taken up via an ABC transporter system that consists of FecA, -B, -C, -D, and -E (Staudenmaier et al. 1989). *E. coli* also has a ferrous iron transport system consisting of polypeptides encoded by the *feoA, -B*, and *-C* genes. The product of the *feoB* gene has a typical ATP-binding motif at the N-terminal end (Kammler et al. 1993). The Sfu system of *Serratia marcescens* is concerned with transport of ferric iron across the membrane (Angerer et al. 1990). The Sfu proteins constitute a typical ABC transporter in which SfuA is localized in the periplasm, SfuB is a cytoplasmic membrane protein, and SfuC is a membrane-bound protein carrying a dinucleotide-binding motif.

Previous studies and an evaluation of the genome sequence of *Synechocystis* sp. PCC 6803 (Boyer et al. 1987; Kaneko et al. 1996; Kotani and Tabata 1998; Straus 1994; Trick and Kerry 1992) revealed the presence of a number of genes with homology to known and well characterized iron transport proteins either of *E. coli* (Braun 1998) or of pathogenic bacteria (Braun 2001; Chen et al. 1993; Ferreiros et al. 1999). These genes and the possible function of their gene products are listed in Table 1.3.

Table 1.3: Summary of *Synechocystis* sp. PCC 6803 genes with homology to genes encoding proteins shown to be involved in iron assimilation and transport in *E. coli* or pathogenic bacteria. The results were obtained using the BlastP program of Cyanobase (http://www.kazusa.or.jp/cyano/cyano.html (Michel and Pistorius 2003).

Synechocystis sp. PCC 6803 ORF's	Homology to gene / species of origin	Function of gene product in *E. coli* or pathogenic bacteria
sll1202 sll1406 sll1409 slr1490	fhuA / *E. coli*	Ferrichrome-binding protein in the outer membrane
sll1202 sll1406 sll1409 slr1490	iutA / *E. coli*	Ferric aerobactin-binding protein in outer membrane
sll1202 slr1491 slr1492 slr1319	fhuD / *E. coli*	Hydroxamate-type siderophore-binding protein in the periplasm
slr1319	fecB / *E. coli*	Ferric dicitrate-binding protein in the periplasm
slr1316	fecC / *E. coli*	Subunit C of ferric iron dicitrate transporter in the cytoplasmic membrane
slr1317	fecD / *E. coli*	Subunit D of ferric iron dicitrate transporter in the cytoplasmic membrane
slr1318	fecE / *E. coli*	ATP-binding protein of ferric iron dicitrate transporter in the cytoplasm
slr1392	feoB / *E. coli*	Protein of ferrous iron uptake system in the cytoplasmic membrane
slr1295 (futA1) slr0513 (futA2)	sfuA / *S. marcescens*	Transferrin-binding protein in the periplasm
slr0237 (futB)	sfuB / *S. marcescens*	Membrane-integral protein of the Sfu-Fe^{3+}-transporter
sll1878 (futC)	hitC / *H. influenzae* sfuC / *S. marcescens*	ATP-binding protein of the Sfu-type Fe^{3+}-transporter

Experimental evidence for a functional role of any of these proteins in iron uptake is still scarce. Interposon mutagenesis experiments showed that a Slr1392-free *Synechocystis* sp. PCC 6803 mutant, a protein homologue to FeoB in *E. Coli*, was impaired in its overall iron uptake capacity (Katoh et al. 2001a). The *feo* system is induced by low intracellular iron concentrations achieved by iron starvation or inactivation of the *fut* system. The genes *slr1295*, *slr0513*, *slr0237*, and *sll1878* of *Synechocystis* sp. PCC 6803 show significant homology to the *fut* genes (ferric iron uptake transporter) of pathogenic bacteria such as, e.g. *Serratia marcescens*, *Neisseria spp.* or *Haemophilus influenzae* (Kirby et al. 1998), and therefore, were named *futA1*, *futA2*, *futB*, and *futC*. The genes of the *fut* system in pathogenic bacteria are functional in iron transport if iron is present as transferrin or lactoferrin encoding an ATP binding cassette (ABC)-type ferric iron transporter. Since the inactivation of *futB* or *futC* or of both *futA1* and *futA2* in *Synechocystis* sp. PCC 6803 greatly reduced the activity of ferric iron uptake and also growth, it was concluded that the *fut* system in *Synechocystis* sp. PCC 6803 is functional in iron acquisition. The *futA1* gene in *Synechocystis* sp. PCC 6803 appears to encode a periplasmic protein, playing a redundant role in iron binding (Katoh et al. 2001a), while FutA2 was mainly detected in the thylakoid membrane copurifying with PS II (Tölle et al. 2002). The deduced products of *futB* and *futC* genes contain nucleotide-binding motifs and belong to the ABC transporter family of inner-membrane-bound and membrane-associated proteins, respectively (Katoh et al. 2001a; Katoh et al. 2001b).

As described in chapter 1.3.2, free intracellular iron leads to the formation of ROS causing oxidative stress and compromising cell components. To prevent damage, the intracellular iron is bound and stored via bacterioferritin (Laulhere et al. 1992) and as shown by Bullerjahn and co-workers (Durham and Bullerjahn 2002) the soluble pool of the DpsA protein in *S. elongatus* PCC 7942 contains detectable iron and has been suggested to function as a ferritin-type protein involved in metal ion homeostasis of the photosynthetic apparatus. Instead of eukaryotic ferritins and bacterioferritins in heterotrophic bacteria, e.g. *E. coli*, there is not much known about bacterioferritins in cyanobacteria. The bacterioferritin from *Synechocystis* sp. PCC 6803 has a molecular mass of 400 kDa and is built up from 19 kDa subunits. Its N-terminal sequence shows 73% identity with that of the *E. coli* bacterioferritin subunit. It contains 2300 atoms of iron and 1500 molecules of phosphate per ferritin molecule and 0.25 haem residues per subunit. In contrast to eukaryotic ferritins, bacterioferritin from *Synechocystis* sp. PCC 6803 is not inducible by iron. It is expressed at constant levels independently of the intracellular iron status, even at very low iron concentration (Laulhere et al. 1992).

1.3.5 Expression of iron-regulated proteins

In *S. elongatus* PCC 7942 two proteins have been identified by biochemical investigations that are highly expressed under iron deficiency and that are shown to be associated with the cytoplasmic membrane. These are IrpA (Reddy et al. 1988) and MapA (Webb et al. 1994).

IrpA (iron regulated protein A) encodes a 38.6 kDa protein with a putative N-terminal signal sequence of 44 amino acid residues for incorporation into the cytoplasmic membrane. An IrpA-free *S. elongatus* PCC 7942 mutant was unable to grow under iron-deficient conditions. Due to its location in the cytoplasmic membrane, IrpA was assumed to function in iron assimilation. So far only one homologous protein has been detected in *Vibrio* sp., which seems to be iron-regulated, since it has a putative Fur box upstream (see chapter 1.3.5.2) of its transcriptional start (Panina et al. 2001). A palindromic sequence of 25 bp length located 20 nucleotides upstream of the suggested transcriptional start was shown to be involved in iron-dependent regulation of *irpA* transcription (see chapter 1.3.5.4) (Durham et al. 2003; Reddy et al. 1988). Recently, it has been shown that in *S. elongatus* PCC 7942 the gene *irpA* is located in an operon with the gene *slr1461* (Nodop et al. 2008). The function of the corresponding protein is so far unknown. Previously, it has already been suggested that *irpA* is located in an operon and that the genes of this operon encode proteins of an iron acquisition system (Reddy et al. 1988).

The *mapA* transcript is not detectable in cultures grown under regular conditions, but is accumulated in cells grown in iron-deficient media. However, the promoter of this gene does not resemble other bacterial iron-regulated promoters. The corresponding 34 kDa MapA protein was found to be associated with the cytoplasmic membrane, but was additionally present in the thylakoid membrane under iron starvation (Webb et al. 1994). The N-terminus of MapA has homology to the inner envelope chloroplast protein E37, while the C-terminal part resembles bacterial iron-binding proteins. The exact function of this protein remains unclear, because like IrpA, MapA has no counterparts in any so far sequenced and annotated cyanobacterial genomes (Nodop et al. 2008).

1.3.5.1 Characteristics and function of the Isi-protein family

The proteins IsiA and IsiB (Isi = iron stress-induced protein) are among the best-characterised proteins shown to be highly expressed under iron deficiency in cyanobacteria (Laudenbach et al. 1988; Laudenbach and Straus 1988).

IsiB (flavodoxin) contains FMN as cofactor. Under iron limitation, it can functionally replace the iron-sulfur protein ferredoxin, which is an electron carrier in a number of reactions (Burnap et al. 1993; Straus 1994). It has been shown that an IsiB-free *Synechocystis* sp. PCC 6803 mutant can grow on iron-deficient medium, suggesting that IsiB is not absolutely required for growth under iron starvation (Kutzki et al. 1998). This is in agreement with results indicating that ferredoxin is never completely lost under iron-limiting growth conditions, even from severely iron-starved *S. elongatus* PCC 7942 cells (Sandström et al. 2002). However, it has been shown that accumulation of flavodoxin contributes to enhanced cyclic electron flow activities around PS I (Hagemann et al. 1999).

IsiA is a hydrophobic protein of about 37 kDa that shares strong homology with PsbC, the CP43 protein of PS II, and is therefore also called CP43' (Burnap et al. 1993; Riethman and Sherman 1988). It has been assigned several functions (Barber et al. 2006; Nield et al. 2003). IsiA is predicted to have six transmembrane helices such as CP43, and since the chlorophyll binding sites (histidine residues) are conserved, IsiA is assumed to bind 12 chlorophylls (Bricker and Frankel 2002). The major difference is that IsiA lacks about 100 amino acids of the large hydrophilic loop that links the luminal ends of helices V and VI of CP43 and therefore, has 342 rather than 472 amino acid residues. Expression of IsiA causes a blue-shift in the room temperature chlorophyll a absorbance peak from 680 to 673 nm, and the chlorophyll a fluorescence at 77K becomes dominated by a high emission at 685 nm (Burnap et al. 1993; Falk et al. 1995; Öquist 1974). Possible functions concerning IsiA were intensively discussed during the last years.

Based on the results of the characterisation of a chlorophyll protein complex called CPVI-4 being expressed under iron limitation in *S. elongatus* PCC 7942 and shown to contain the IsiA protein (Burnap et al. 1993), it was proposed that IsiA acts as a chlorophyll store under iron limitation and aids the recovery of PS II and PS I when iron again becomes available in the environment (Riethman and Sherman 1988).

It was shown for *Synechocystis* sp. PCC 6803 (Bibby et al. 2001b; Bibby et al. 2001c) and for *S. elongatus* PCC 7942 (Boekema et al. 2001) that IsiA can form an additional membrane-integral light-harvesting antenna around trimeric PS I complexes. Spectroscopic evidence confirmed a possible function as a ring like functional antenna for PS I composed of 18 IsiA molecules, directing more excitation energy to PS I under iron deficiency (see Figure 1.13) (Andrizhiyevskaya et al. 2002; Melkozernov et al. 2003; Nield et al. 2003). The PS I-IsiA supercomplex allows *S. elongatus* PCC 7942 to maintain cyclic electron flow around

PS I as well as respiratory electron flow to compensate for the loss of PS II activity as a cause of prolonged iron depletion. For *Prochlorococcus marinus* strain SS120 it was also shown that an antenna ring around PS I consisting of chlorophyll a/b-binding Pcb proteins is present (Bibby et al. 2001a).

IsiA can also interact with PS I monomers forming single or double rings with multiple copies of IsiA (see Figure 1.13). The compensation depends on the phase of iron starvation (Kouril et al. 2005). It has been shown that after a prolonged growth of *Synechocystis* sp. PCC 6803, the largest complexes bind 12-14 units in an inner ring and 19-21 units in an outer ring around a PS I monomer. Fluorescence excitation spectra indicated an efficient light harvesting function for all PS I-bound chlorophylls. Additionally, in *Synechocystis* sp. PCC 6803 a significant part of IsiA builds supercomplexes without PS I forming single or double rings, which can be closed or incomplete (Yeremenko et al. 2004).

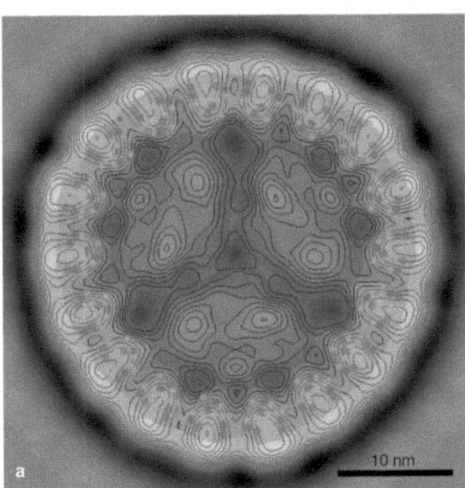
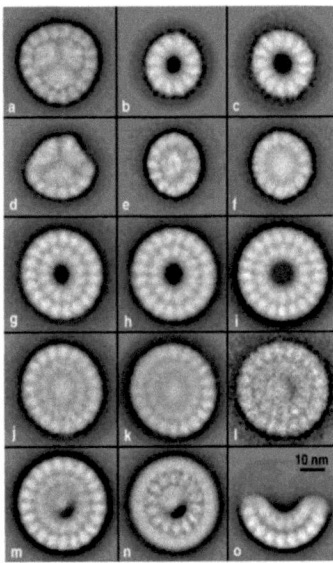

Figure 1.13: Processed top views of PS I-IsiA and IsiA supercomplexes obtained by electron microscopy. (a) left figure shows a coloured image of a PS I trimer and 18 IsiA copies, each containing 12 chlorophyll molecules (Boekema et al. 2001). Closed rings of IsiA consist of 12, 13, or 18 copies in a single ring (b, c, and a, respectively) along with 19, 20, and 21 copies in a second ring surrounding 12, 13, and 14 copies in an inner ring (g-i). The centre of the complexes can be occupied by a either a PS I trimer (a) or a monomer (e, f, and j-l). The images of panels m and n contain a larger number of projections than image l but have the same number of 14 + 21 IsiA copies. A specific complex of trimeric PSI with seven IsiA proteins attached (d) (Yeremenko et al. 2004).

These kind of IsiA complexes could act as chlorophyll sinks (Burnap et al. 1993) to prevent high quantities of unbound potentially hazardous chlorophyll. Free IsiA complexes may also play a role in shading, removing excess light excitation energy from the membranes and thus preventing overexcitation and photodamage of PS II and PS I (Ihalainen et al. 2005; Ivanov et al. 2006; Ivanov et al. 2000; Sandström et al. 2002; Sandström et al. 2001). It has been shown, that mutations in the PS I subunits PsaF/J and PsaL decrease the binding affinity of

IsiA and generate PS I-IsiA particles with incomplete IsiA rings (see Figure 1.13). Thus, these subunits facilitate the formation of closed IsiA rings around PS I but are not obligatory components in the formation of the PS I-IsiA supercomplexes.

1.3.5.2 Regulation of *isiA* gene expression

In several cyanobacteria the genes *isiA* and *isiB* compose an operon structure, generating a mono and dicistronic message of *isiA* (*isiA* and *isiAB*) (Laudenbach et al. 1988). Elevated transcription of the *isiAB* operon is seen under iron limitation (Burnap et al. 1993; Laudenbach et al. 1988), under heat stress (Fulda et al. 2006) or salt stress (Geiss et al. 2001; Vinnemeier and Hagemann 1999; Vinnemeier et al. 1998), under high light stress (Havaux et al. 2005),and in some cyanobacterial mutants, such as a Cyt c_6-deficient (Ardelean et al. 2002) and a PsaJF-free *Synechocystis* sp. PCC 6803 mutant, respectively (Jeanjean et al. 2003). Moreover, recent results show that *isiA* and *isiB* in *Synechocystis* sp. PCC 6803 (Li et al. 2003; Murata 2003) and in *S. elongatus* PCC 7942 (Michel et al. 2003; Singh et al. 2004; Yousef et al. 2003) are highly expressed under peroxide stress. In *Synechocystis* sp. PCC 6803 IsiA is also expressed in significant amounts in the stationary growth phase (Singh and Sherman 2006). Either multiple input signals induce expression of IsiA and IsiB, or all these stress conditions lead to oxidative stress. The redox poise affects the photosystems and with it the redox state of the other photosynthetic components. It is suggested that IsiA expression depends in parts on the redox state of the PQ-pool (Pietsch et al. 2007). Therefore, it could be assumed that oxidative stress is the superior signal for expression of the *isiAB* operon.

The gene *isiA* is transcriptionally repressed via a cyanobacterial Fur-homologue protein (Ferric uptake repressor) as shown for *S. elongatus* PCC 7942 (Ghassemian and Straus 1996). Fur, being first identified in *Salmonella typhimurium* and in *E. coli* (Ernst et al. 1978; Hantke 1981), is a small 17 kDa protein, which belongs to the family of Crp/Fnr transcriptional regulators. In the presence of its co-repressor Fe^{2+} it binds in a dimeric state to specific operator sequences on the chromosome, called Fur boxes, upstream of the corresponding genes and represses their transcription (Escolar et al. 1999; Hantke and Braun 1998). When the intracellular Fe^{2+} concentration drops below a specific threshold, the repressor complex physically disassembles and allows transcription of the downstream located genes. Apart from the *isiAB* operon no other cyanobacterial genes have yet been shown to be regulated via Fur, although a number of Fur boxes have been postulated upstream of cyanobacterial genes (Straus 1994). Attempts to insertionally inactivate the *fur* gene generated merodiploid *S. elongatus* PCC 7942 mutants (Ghassemian and Straus 1996; Michel et al. 2001). The *Synechocystis* sp. PCC 6803 genome sequence, however, reveals the presence of three fur-homologue genes, *sll0567*, *sll1937*, and *slr1738*. On the basis of results of Kunert (Kunert 2001) Sll0567 is probably the Fur-orthologue in *Synechocystis* sp. PCC 6803 (insertional inactivation of *fur* always generated merodiploid mutants). In addition to the Fur box, an additional nucleotide sequence was suggested to be also involved in transcriptional regulation of the *isiAB* operon (Kunert 2001; Kunert et al. 2003). Moreover Slr1738 has also been suggested to be involved in regulation of the *isiAB* operon of *Synechocystis* sp. PCC 6803 (Li et al. 2003).

In *Synechocystis* sp. PCC 6803 the steady state *isiA* mRNA pool is posttranscriptionally regulated. Recently it has been shown, that the IsiA expression is controlled by iron stress-repressed RNA (IsrR), a cis-encoded antisense RNA transcribed from the *isiA* non-coding strand. This RNA contains short and only partial antisense complementary to the *isiA* sequence, forming a unique double structure with the target transcript. This structure affects mRNA stability and translation causing the regulation of gene expression this way. Artificial overexpression of IsrR under iron stress causes a strongly diminished number of IsiA-PS I supercomplexes, whereas IsrR depletion results in premature expression of IsiA (Duhring et al. 2006).

1.3.5.3 The IdiA protein

IdiA (iron deficiency induced protein A) is a protein of 35 kDa, whose expression is highly increased under iron-deficient growth conditions in *S. elongatus* PCC 6301 and PCC 7942 (Michel 1996; Michel and Pistorius 1992) and in the thermophilic species *Thermosynechococcus elongatus* BP-1 (Exss-Sonne 2000; Exss-Sonne et al. 2000; Michel et al. 1998). In contrast to IsiA, the IdiA protein becomes expressed in highly elevated amounts during the early phase of iron starvation (Tölle et al. 2002). In addition, the IdiA expression is slightly elevated under Ca^{2+} or Mg^{2+} deficiency, and when cells were cultivated with ammonium as sole N-source (Michel 1996).

Biochemical fractionation methods and immunocytochemical techniques showed that IdiA is mainly located intracellularly and preferentially associated with the cytoplasmic side of the thylakoid membranes. Moreover, blue native gel electrophoresis and ion exchange chromatography provided evidence that IdiA in part co-purifies with PS II complexes (Exss-Sonne et al. 2000). Characterization of an IdiA-free *S. elongatus* PCC 7942 mutant revealed that in the absence of IdiA the growth rate was severely reduced in the mutant compared to wild type, and that PS II was mainly damaged, while PS I was not affected. This damage of PS II was already seen even under sufficient growth conditions and increased under iron-deficient growth conditions. Further experiments have given evidence that IdiA is a protein, which protects and shields the acceptor side of PS II against oxidative damage (see below). A model of the function of IdiA is given in Figure 1.14.

The acceptor side of PS II becomes progressively exposed towards the cytoplasm in the cause of ongoing iron starvation induced partial degradation of the phycobilisomes. Thus, a protection of the here located redox active components, like the non-heme iron on the D1/D2 heterodimer, have to be protected against oxidative damage especially under iron deficient conditions. However, the exact mechanism of this protection has remained unclear (Exss-Sonne et al. 2000). In agreement with a PS II acceptor side function of IdiA, is the observation that in *S. elongatus* PCC 7942 the inhibitory effect of the herbicide bentazone, which binds mainly to the acceptor side of PS II, is significantly reduced when the IdiA protein is expressed (Bagchi et al. 2003).

Introduction

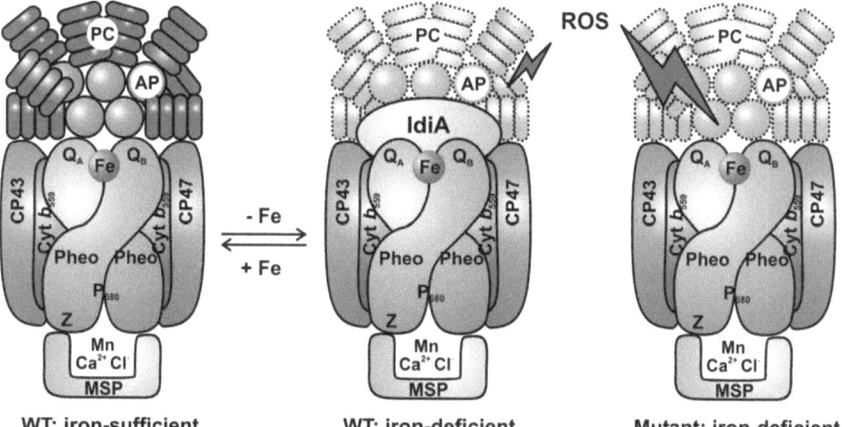

| WT: iron-sufficient | WT: iron-deficient | Mutant: iron-deficient |

Figure 1.14: Model of the protective function of IdiA for PS II in *S. elongatus* PCC 6301/PCC 7942 under iron-limiting growth conditions. The reduction of phycobilisomes in number or size associated with PS II is given in a light blue colour. Whereas the WT expresses the IdiA protein, which is suggested to protect the acceptor side of PS II against oxidative stress, an IdiA-free mutant is more prone to damage by ROS under iron starvation (Michel & Pistorius 2003).

The protective function of IdiA for PS II has recently gained further support, as IdiA has been detected in highly purified PS II complexes from iron starved *Thermosynechococcus elongatus* BP-1 cells (Lax et al. 2007). A comparison of PS II electron microscopy projection maps with PS II high resolution X-ray structures, shown in Figure 1.15, confirms the binding of IdiA on the cytoplasmic side of PS II.

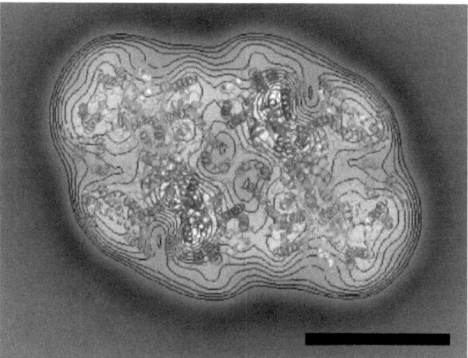

Figure 1.15: Comparison of the PS II electron microscopy projection map with the PS II high-resolution X-ray structure. The PS II single particle is displayed from the lumenal side with CP43 (yellow), CP47 (purple), D1 (blue) and D2 (pink). Other membrane-bound subunits have been indicated in olive-green; PsbZ and extrinsic subunits have been omitted. IdiA is positioned within the bright green line at the cytoplasmic side. The space bar represents 10 nm (Lax et al. 2007).

When the time course of expression of *idiA* and *isiA* mRNA was investigated (see Figure 1.16), it became obvious that the *idiA* mRNA was detected after 6 h reaching a maximum after about 24 h, while the *isiA* mRNA was firstly detected after 24 h reaching a maximum

25

after about 60 h of growth in iron-depleted medium under the chosen growth conditions (Michel and Pistorius 2003; Yousef et al. 2003). This implies that shortly after the onset of iron-deficient growth conditions the more labile PS II is modified by the IdiA protein protecting the non-heme iron of the D1/D2 heterodimer. This modification provides an efficient linear electron transport between PS II and PS I during the initial phase of iron deficiency. After prolonged iron starvation when damage of PS II can not any longer be effectively prevented, modification of PS I by a newly synthesized chlorophyll a antenna consisting of IsiA molecules takes place providing an efficient cyclic electron transport around PS I.

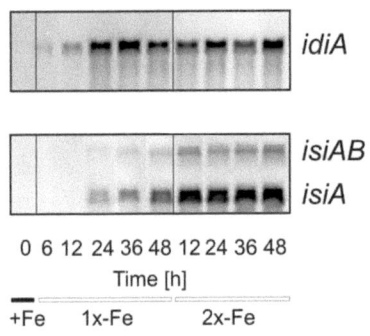

Figure 1.16: Time course of *idiA* and *isiA* mRNA transcription in *S. elongatus* PCC 7942 cells cultivated under iron-limiting growth conditions (Michel and Pistorius 2003).

1.3.5.4 Regulation of *idiA* expression

Using a reverse genetic approach with gene-specific oligonucleotides, the gene encoding IdiA in *S. elongatus* PCC 6301 (Michel et al. 1996) and its genomic region was identified and sequenced (EMBL database entry Z48754; (Michel et al. 1999)). *IdiA* from *S. elongatus* PCC 7942 has also been sequenced (EMBL database entry AJ319672; (Tölle et al. 2002)). Southern blot analysis revealed that *idiA* in *S. elongatus* PCC 6301 and PCC 7942 exists as a single copy gene in the chromosome (Michel et al. 1999). It consists of 1101 bp encoding a protein of 39.2 kDa. IdiA is synthesized as a precursor and cleaved between two alanine residues in position 36 and 37 to its mature form of 35.2 kDa.

Sequencing of the 5.8 kb *Hind*III fragment of genomic DNA from *S. elongatus* PCC 6301 carrying the IdiA gene, resulted in the identification of a novel operon structure located downstream of *idiA*. This operon contains three open reading frames, which are transcribed in an opposite direction to *idiA*. The first gene, called *idiB*, is located immediately downstream of *idiA* (see Figure 1.17).

IdiB encodes a putative helix-turn-helix transcriptional regulator (Michel et al. 1999; Michel et al. 2001) of the Crp/Fnr family (Holm et al. 1994; Unden and Guest 1985). The gene *idiB* is transcribed in opposite direction to *idiA* and has strong similarity to other well-characterised transcriptional regulators of *S. elongatus* PCC 7942, such as NtcA (Frias et al. 1993; Herrero et al. 2001) and CysR (Nicholson and Laudenbach 1995). Further experiments showed that the *idiA* promoter consists of an atypical -10, an atypical -35 region, and a 14 bp long palindromic sequence 4 bp upstream of the -35 region. These structures resemble the binding site of Fnr/Crp-type helix-turn-helix transcription factors. The absence of this palindromic sequence or a 3 bp mutation in a putative -10 region abolished promoter activity completely. The transcriptional start point has been predetermined to position bp 2126 on Z48754 (Lim 2003).

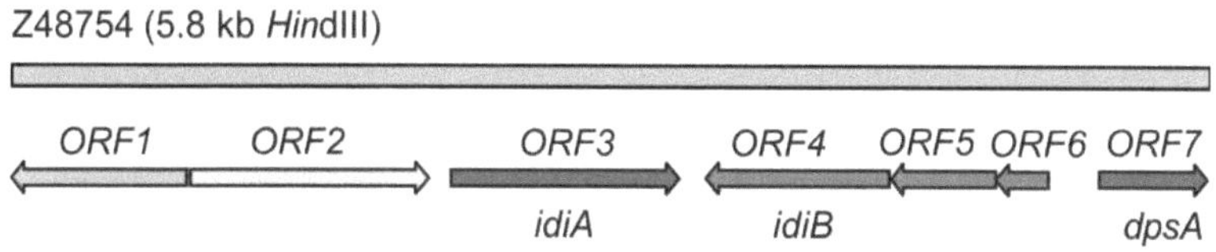

Z48754 (5.8 kb HindIII)

ORF1	ORF2	ORF3	ORF4	ORF5 ORF6	ORF7
		idiA	idiB		dpsA

Figure 1.17: Physical map of EMBL database entry Z48754. The 5.8 kb HindIII fragment of genomic DNA from S. elongatus PCC 6301 carries the idiA gene and the genes idiB and dpsA, whose gene products regulate and influence transcription of idiA (Michel et al. 1999).

Using mobility shift assays, it was shown that 6His-tagged IdiB bound to a 59-bp fragment of the idiA regulatory region that included the palindrome (Michel et al. 2001). Moreover, an IdiB-free S. elongatus PCC 7942 mutant exhibited no IdiA expression implying that IdiB is a positively acting transcriptional factor regulating idiA mRNA expression in S. elongatus PCC 7942 (Michel et al. 1999). In addition to IdiB the DpsA protein encoded by the dpsA gene, which is located downstream of idiB on the S. elongatus PCC 6301 and PCC 7942 chromosome, also had an effect on IdiA expression, although most likely in an indirect way (Michel et al. 2003; Michel et al. 1999).

The genes irpA as well as mapA contained also putative operator sequences, containing the binding site for helix-turn-helix transcriptions factors, but a transcriptional analysis reveals that only irpA is regulated by IdiB. Immuno blot analyses of an idiB knock-out mutant revealed also a strongly decreased IrpA content under iron deficient conditions.

The transcription of the gene encoding IdiB is also increased under iron limitation (Michel et al. 2003; Yousef et al. 2003). IdiB is probably not controlled by its own promoter (Lim 2003) because in S. elongatus PCC 7942 the primary tricistronic message together with ORF5 and ORF6 was found as well as the secondary processed dicistronic transcript with ORF5 and the monocistronic idiB transcript, shown in Figure 1.18 (Yousef et al. 2003).

Introduction

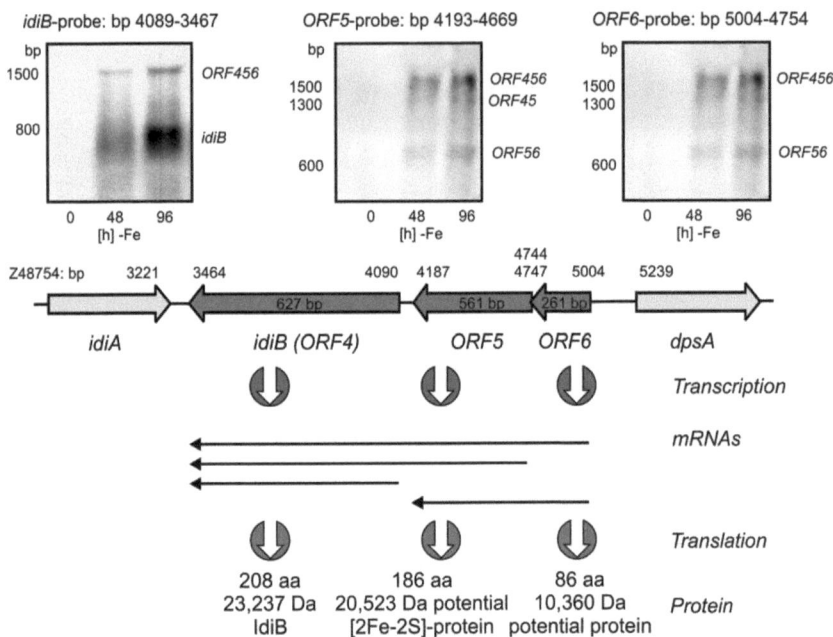

Figure 1.18: Transcriptional analyses of the IdiB operon in S. elongatus PCC 7942 and its primary and secondary transcripts as well as the basics of the putative encoded proteins (Yousef et al. 2003).

So far, the transcription factor (or factors) regulating the expression of this operon has not been identified. The analysis of the deduced amino acid sequence of *ORF5* revealed that it encodes a protein with similarity to [2Fe-2S] cluster-containing proteins (Pietsch et al. 2007). The fact that the expression of a potential [Fe-S]-protein is increased under iron starvation came rather unexpected, since it is widely accepted that some [Fe-S]-proteins like ferredoxins are replaced by flavodoxins under iron-depleted conditions (Smillie 1965). In this work it will be shown that the expression of the protein encoded by *ORF5* is iron-dependently regulated and that this protein indeed contains an iron-cofactor. The protein encoded by *ORF6* reveals similarities to fur transcription factors.

Since IsiA (and IsiB) expression is regulated via the transcriptional repressor Fur and IdiA expression is regulated by IdiB, these results imply that IdiA and IsiA expression are controlled by different regulatory pathways (Yousef et al. 2003). However, certain results suggest that a cross-talk between these two transcriptional regulatory pathways exists. In a *fur*-merodiploid S. elongatus PCC 7942 mutant, a partial expression of IdiA is observed already under iron-sufficient conditions (Michel et al. 2001). On the other hand, in an IdiB-free S. elongatus PCC 7942 mutant, the increase in 77 K fluorescence at 685 nm due to IsiA expression is much less than in wild type, although an *isiA* expression compared to the wild type can be observed (Yousef et al. 2003). This suggests that either the structure or the assembly of the PS I-IsiA complex is affected when IdiB-mediated transcription is absent. It is also possible that IsiA in the IdiB-free mutant has partially a different function, e.g. being involved in energy dissipation as suggested by Sandström et al. (Sandström et al. 2002). The context of *idiA* and *isiAB* regulation is combined in the following model shown in Figure 1.19.

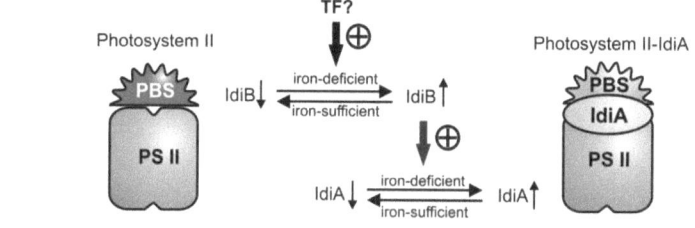

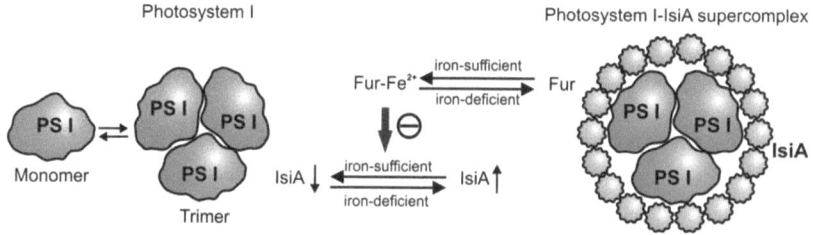

Figure 1.19: Model of transcriptional regulation of IdiA (Michel et al. 2001) and IsiA expression (Ghassemian and Straus 1996). Abbreviations are given in previous figures and tables.

1.3.6 Summary of the modification of PS I by IsiA and PS II by IdiA

Due to their oxygenic photosynthetic life-style, cyanobacteria have a higher demand for iron than other heterotrophically growing eubacteria, because the photosynthetic multi-protein complexes, involved in electron transport from water to ferredoxin, possess a fairly high number of iron atoms as cofactors. Furthermore, due to the generation of oxygen, the interrelationship and complexity of iron homeostasis and oxidative stress is tighter than in non-photosynthetic eubacteria. Iron limitation greatly affects especially PS II and PS I resulting in an increased production of toxic ROS.

Since their demand for iron and the complex and intertwined problems of iron use, iron availability and toxicity, cyanobacteria developed adaptational mechanisms to prevent iron starvation and the directly associated oxidative stress. These mechanisms include protection and/or modification of the photosynthetic electron transport chain and photosynthetic protein complexes via newly synthesized proteins under iron deficiency. The adaptation of the multi-protein complexes PS II and PS I to iron starvation is a sequential process, which includes the enhanced expression of two major iron-regulated proteins: IdiA and IsiA (Michel and Pistorius 2004). It is suggested that IdiA protects the acceptor side of PS II against oxidative stress under conditions of mild iron limitation in a yet unclear way, whereas prolonged iron deficiency leads to the synthesis of a Chl *a* antenna around PS I-trimers consisting of IsiA protein molecules (Boekema et al. 2001; Michel and Pistorius 2004). The thylakoid membrane of cyanobacteria contains the components of the photosynthetic and the respiratory electron transport chain. Since these two chains intersect and in parts utilise the same redox components, alterations in the activity of the two photosystems will also have an influence on the respiratory activity. The physiological consequences of the alterations occurring under prolonged iron starvation in *S. elongatus* PCC 7942 as described above, are a reduction of linear electron transport activity through PS II and an increase of cyclic electron flow activity around PS I as well as an increase in respiration shown in Figure 1.20 (Michel et al. 2003).

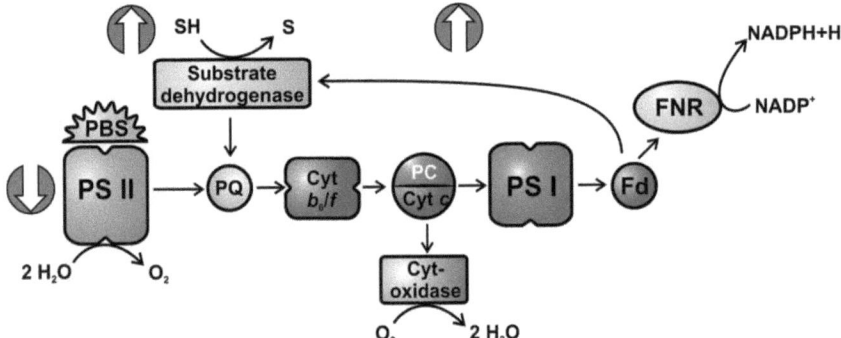

Figure 1.20: Schematic presentation of the changes in relative activities of the electron transport chains in cyanobacteria as a consequence of the adaptation to prolonged iron-limiting growth conditions. Boxed arrows indicate a decline in open chain electron transport and an increase in photosynthetic cyclic electron transport around PS I and respiratory electron transport activity (Michel 2003).

1.4 Aim of research

Cyanobacteria evolved approximately 3.5 billion years ago and were among the first organisms to carry out an oxygenic photosynthetic metabolism. Thus, cyanobacteria were among the first organisms, which had to cope with iron limitation and which had to overcome the problems of iron homeostasis in general. Although iron is an abundant element of the earth's crust, its biological availability is rather poor. Moreover, not only iron starvation is hazardous, but excess of iron is known to cause severe problems, because free Fe^{2+} ions have the potential to produce toxic reactive oxygen species. Since iron is a frequently occurring nutrient limitation in otherwise nutrient-rich habitats, cyanobacteria have evolved sophisticated adaptational strategies to keep their iron homeostasis in balance.

This strategy comprises an optimisation of iron assimilation systems as well as a partial substitution of iron-containing enzymes such a ferredoxins, e.g. by flavin-containing enzymes such as flavodoxin. Moreover, the acclimation to iron starvation comprises the modification of the electron transport chain leading to a reduction of photosynthetic linear electron flow and an increase in photosynthetic cyclic and respiratory electron transport activity.

Although some of the proteins remodelling the electron transport chain have already been identified and partially characterised, e.g. in S. elongatus PCC 7942, where IdiA modifies PS II under early iron starvation and IsiA modifies PS I under prolonged iron starvation, further proteins may participate in the transition of the electron transport chain to iron-deficient growth conditions. One such protein might be IdiC from S. elongatus PCC 7942. The gene encoding IdiC precedes the gene idiB, encoding the transcriptional regulator of IdiA expression, in the so called idiB operon together with ORF6. Intriguingly, the iron starvation induced idiC gene encodes a protein of the thioredoxin-like ferredoxin family with a putative [2Fe-2S] binding site.

One major goal of my thesis was the identification of new iron-regulated proteins, which contribute to the modification of the electron transport chain and a detailed analysis of the function and the characteristics of IdiC protein in S. elongatus PCC 7942.

Another major goal was related to the transcriptional regulation of those proteins that facilitate the modification of the electron transport chain under iron starvation. It is well known that IsiA regulation is mediated by the transcriptional repressor Fur, and that regulation of IdiA expression is facilitated by the transcriptional regulator IdiB. Although there is evidence that both transcriptional regulators act independently of each other, there is also substantial evidence that both regulatory pathways may be intertwined and may share similar input devices at a certain level (Bagchi et al. 2007; Bagchi et al. 2003; Yousef et al. 2003). This question is particularly interesting, since the transcriptional regulator for the iron-regulated idiB operon is still unknown, and since there is a possibility that both regulatory pathways, the Fur as well as the IdiB pathway, are influenced by a yet unidentified ROS- or metal homeostasis-sensing transcriptional regulator, e.g. of the PerR-type, which senses oxidative stress as a mean of unbalanced metal ion homeostasis (Yousef et al. 2003). Considering the details of previous results of our group, another major goal of my thesis was to identify new transcriptional regulators mediating the regulation of the acclimation to iron starvation in *S. elongatus* PCC 7942.

2 Materials and Methods

(A) General information

2.1 Chemicals

Chemicals used in this work were obtained from Roth (Karlsruhe, Germany), Sigma-Aldrich, Fluka (both Munich, Germany), Merck (Darmstadt, Germany) or BioRad (Munich, Germany). Antibiotics were purchased from Serva (Heidelberg, Germany) and Roth. They were stored as 1000-fold concentrated stock solutions at -20 °C.

2.2 Enzymes and kits

Restriction enzymes were purchased from New England Biolabs (Frankfurt a. M., Germany), MBI Fermentas (St. Leon-Rot, Germany) and Promega (Mannheim, Germany).

DNA- and RNA-ladders used in this work were obtained from Peqlab (Erlangen, Germany). Prestained protein ladder was purchased from MBI Fermentas.

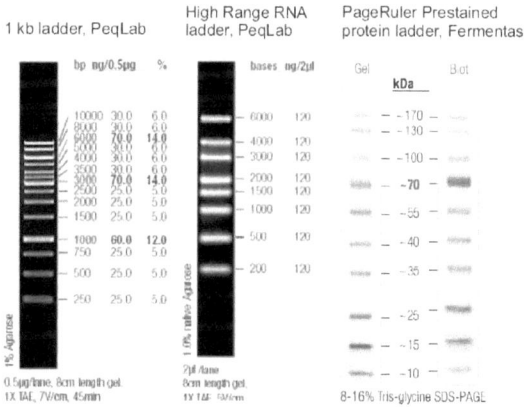

Figure 2.1: Markers used in this work.

The Pwo polymerase was obtained from PeqLab. Phusion™ polymerase was obtained from New England Biolabs. T4 DNA ligase, T4 DNA polymerase and Klenow polymerase were purchased from Promega, Biotherm™ Taq polymerase from GeneCraft (Ares Bioscience GmbH, Cologne, Germany).

Isolation of total RNA from *S. elongatus* PCC 7942 was performed using the RNeasy Mini-Kit manufactured by Qiagen (Hilden, Germany). DNA probes for transcript profiling were polished using QIAquick PCR Purification Kit (Qiagen). DNA was extracted from gels using the QIAquick Gel Extraction Kit (Qiagen). After cleavage with restriction enzymes, DNA fragments were cleaned up using QIAquick Nucleotide Removal Kit or QIAquick PCR

Purification Kit (Qiagen). Plasmids were isolated with the peqGOLD Plasmid Miniprep Kit manufactured by PeqLab.

2.3 Bacterial strains and plasmids

2.3.1 Bacterial strains

The following bacterial strains were used in this work:

Table 2.1: Bacterial and cyanobacterial strains, which were used.

Strain	Genotype/Phenotype	Reference
Escherichia coli DH10B	F-, mcrAΔ(mrr hsdRMS mcrBC), ∅80lacZΔM15 ΔlacX74 deoR recA1 endA1 araΔ139 Δ (ara, leu) 7679, galU, galK, λ- rpsL (StrR), nupG	Invitrogen, Karlsruhe, Germany
Escherichia coli DH5αmcr⁻	SupE sbcB15 hsdR4 rpsL thi Δ(lac-proAB) F´[traD36 proAB⁺ lacI^q lacZ ΔM15] mcr⁻	(Hanahan and Meselson 1983)
Escherichia coli BL21(DE3)	F⁻ ompT hsdS$_B$ (r$_B^-$ m$_B^-$) gal dcm (DE3)	Novagen, Nottingham, UK
Escherichia coli Top10	F- mcrA Δ(mrr-hsdRMS-mcrBC) φ80lacZΔM15 ΔlacX74 recA1 araD139 Δ(ara-leu)7697 galU galK rpsL endA1 nupG	Invitrogen, Karlsruhe, Germany
Synechococcus elongatus PCC 7942	wild type (WT)	Institute Pasteur: Collection Nationale de Cultures de Microorganismes
Synechococcus elongatus PCC 7942 Mutant MuD	idiC merodiploid, Sp^R	Pietsch et al. 2007
Synechococcus elongatus PCC 7942 Mutant K10	Δ idiB, Sp^R	Michel et al. 1999
Synechococcus elongatus PCC 7942 Mutant K9 #1	Δ furII, Sp^R	this work
Synechococcus elongatus PCC 7942 Mutant 2#1	Δ merR, Sp^R	this work

2.3.2 Plasmids

Further information concerning construction and usage of the plasmids listed in table 2.2 is given in the results section and the appendix.

Table 2.2: Plasmids applied or constructed during this work.

Plasmid	Characteristics	Reference
pKPM24	pSVB30 derivative Ap^R, *lacZ*, 5,8 kb *Hin*dIII insert of genomic DNA from *Synechococcus elongatus* PCC 6301	(Michel 1996)
pQE-82L	pT5, *lacO*, *lacI*, RBS, MCS, 6xHis-tag, Ap^R	(Qiagen, Hilden, Germany)
pQE82L-*idiC*	pQE-82L derivative, *lacI*, Ap^R, 593 bp *idiC* insert from *Synechococcus elongatus* PCC 7942	this work
pET-32a(+)	pT7, lacI, MCS, 6xHis-Tag, S-Tag, Trx-Tag, Ap^R	Novagen, Nottingham, UK
pET32a-*idiC*	pET-32a-c(+) derivative *lacI*, Ap^R, with the *idiC* insert from *Synechococcus elongatus* PCC 7942	this work
pUC19	pMB1 origin, MCS, *lacZ*α, Ap^R	(Yanisch-Perron et al. 1985)
pHP45ΩSp^R	Tn5 Sp^R, Ap^R, *lacZ*	(Fellay et al. 1987)
pUC19-*furII*	pUC19 derivative, *lacZ*α, Ap^R, 2,3 kb insert of genomic DNA from *Synechococcus elongatus* PCC 7942 including the *furII* sequence	this work
pUC19-*furII*-Sp^R	pUC19, pHP45ΩSp^R derivative, Ap^R, Sp^R, 2 kb Sp^R cassette disrupting the *furII* gene (AP008231.1, EMBL)	this work
pAM2255	Ptrc, *lacI*q, MCS, Ap^R, NS 2	(Andersson et al. 2000) Texas AM strain file, Prof. PhD. Susan Golden, Texas A&M, College Station, Texas, USA
pAM2314	Ap^R, Sp^R, MCS, NS 1	(Andersson et al. 2000) Texas AM strain file Prof. PhD. Susan Golden, Texas A&M, College Station, Texas, USA
p39	pAM2255, pAM2314 derivative, Ptrc, *lacI*q, MCS, Sp^R, NS1	Jan Schöning 2004, Molecular Cellphysiology, University of Bielefeld, Bielefeld, Germany
p39-*idiChis6*	p39 derivative, Ptrc, *lacI*q, Sp^R, NS1, 685 bp *idiChis6* insert from pQE82L-*idiC*	this work
p8S4-G1	pMCL200 derivative, Cm^R, Km^R; 8 kb insert of genomic DNA from *Synechococcus elongatus* PCC 7942 including the *merR* sequence and 1,2 kb of transposon DNA	Prof. PhD. Susan Golden, 2007 Texas A&M, College Station, Texas, USA

2.3.3 Oligonucleotides

The following primers were used to generate Dig-dUTP-labelled gene specific DNA probes for transcript analyses. The primers were designed with the software *Primer Designer 2.2* (Scientific and Educational Software, 1994) and purchased from Sigma-Aldrich (Munich, Germany).

Table 2.3: Primers used for preparation of DNA probes for RNA-DNA hybridization, part A.

Primer	EMBL Accession number	Amplified product	DNA sequence 5`→3` direction
idiC F idC R	i*-Z48754	576 bps	CAAGGATCCAACTGCCGTTCTA TCGAAGCTTTTTAGCCCACGGC
idiA F idiA R	Z48754	1082 bps	GTTGACATGTCTGAATCAATGTTCAGT CTGTGATCAGTCTTATTTCCAGCCAC
idiB F idiB R	Z48754	623 bps	TGATTGCCAGTCACGTAACC GGCATCTATGGCATCAATCG
isiA F isiA R	M23639	928 bps	CACTCTCCTTCATGCAGACT GCATTGGATAGCCAAGCACG
isiB F isiB R	M23639	506 bps	AAATTGGCTTATTCTACGGA TTAAAGACCGAACTCTGACT
isiC F isiC R	CP000100.1	514 bps	TGTTGGCTAACTTGCTCGGG AAGCGGCTTCTCTGGATACA
irpA F irpA R	M22058	1056 bps	TGATCGTGACAGGCTCTCAG GGAGAACTTGACGCTCTAGG
irpB F irpB R	AP00823.1	507 bps	GGCTGCTGAACATCGACAGG TCGTGATTGCGATCCTCTGG
mapA F mapA R	L01621	839 bps	ACTACCGGATTCAGTCAAGC CCTGAATGCCGGTCTTGTCT
dpsA F dpsA R	U19762	514 bps	CTGCCATGACGAATACAGGT GGAGCCAAGAAGTGAGCAA
rnpB F rnpB R	AF056387	312 bps	GAAAGTCCGGGCTCCCAAAAGA TAAGCCGGGTTCTGTTCTCTGTC

*inverted sequence Z48754

Table 2.4: Primers used for construction of plasmids listed in Table 2.2.

Primer	EMBL Accession number	Amplified product	DNA sequence 5`→3` direction
AMO 541 idiB-F9	Z48754	3532 bps	CTCGCGCCTTAGCTCAGACTT ATCCTTCACCGAACCATAGC
IdiB-F2 idiB-R3	Z48754	692 bps	GCCAGTGGATTCAACGCGAT GGGTCTAGAGTGAGAGCAATGAG
orf5 F orf5 R	i-Z48754	576 bps	CAAGGATCCAACTGCCGTTCTA TCGAAGCTTTTTAGCCCACGGC
orf5his6 F orf5his6 R	-	594 bps	ATGAGAGGATCTCACCATCA TTAGCCCACGGCAGTGCCTG
furII F furII R	AP008231.1	2322 bps	CATGCCAGTGCTCGTCGTCT CGAGCTGAATCCAGTGCGAA
merR F merR R	CP000100.1	2042 bps	TAGGTTGCTGCAATCCGGTC TTCTATGACGAGCATGCTTC

Table 2.5: Primers used for preparation of DNA probes for RNA-DNA hybridization, part B.

Primer	JGI gene name	Amplified product	DNA sequence 5'→3' direction
ackA F ackA R	slr2079	419 bps	GAGCAGATGGAGCAGTTGTT TGCTGATTCAGGACCGTGCT
ftr1 F ftr1 R	slr2421	437 bps	CTACGCCTACTCCTAGAAGC CCTTGGCTGAGTTGATTGAG
ftrC F ftrC R	slr1734	358 bps	GACCCAGACGACCAGCCCAG AGCAGGCTGGGTCGTGGCAC
futA F futA R	slr1409	476 bps	AATCATCCGTGCTCACCGCT AACGCCGGCACCACTGACAT
futB F futB R	slr1407	502 bps	GGATGGCCTTGGCGTTGATG CGCACCGCTAGCTGAATTGG
futC F futC R	slr1406	468 bps	CACTGTTTCCGCATCTAACG GATACTCTGTGGCCAAGAAG
irpA F irpA R	slr1462	469 bps	GCAATGGACACCTGGCCTAT GCTGACCGCGATAGACATTC
irpB F irpB R	slr1461	484 bps	CCTCGATCCAAGCATTCAAG CTGTTCTGAGAGTGCCGCGA
mapA F mapA R	slr1408	429 bps	ATCATCCTACGATCGCCGCT TGCAACCATCGATCTAGGAG
merR F merR R	slr1739	410 bps	GACCGCTACACTGCTCAAAA TGGTTGCAATTAACGGACAG
rnpB F rnpB R	slr1148	312 bps	GAAAGTCCGGGCTCCCAAAAGA TAAGCCGGGTTCTGTTCTCTGTC
nat F nat R	slr0650	326 bps	TGATTCGAGAGTTGGCCAGC CCGTTGATTCCAGTCGAGCA
isiC F isiC R	slr1540	541 bps	TGTTGGCTAACTTGCTCGGG AAGCGGCTTCTCTGGATACA
isiB F isiB R	slr1541	389 bps	CTGACTTTAGCTGGCTGACC TGCGAATGCTGATGCCAGTG
isiA F isiA R	slr1542	486 bps	ACACCACTTGCTGTTCCTCG TTGGATAGCCAAGCACGAGG
ORF6 F ORF6 R	slr2172	257 bps	GACTCTCGGCACTCTTCGCA CATCGTTGTTGCTCGTGGTA
pgam F pgam R	slr2078	521 bps	AACGACCGCTAGCCGTGAAG AAGACAGGATGGAGAGCGAG
somA2 F somA2 R	slr1607	456 bps	CGGTAATGAGGTCAACAGCA ACTTGGTCAGCACCTCGGTT
somb1 F somb1 R	slr1463	521 bps	CGTTCCAACAGCATCACCTC TTCCGGTAAGCGTAGGCCAG
sufB F sufB R	slr1140	471 bps	AAATACGGCTTCGTCACGGA CACACTGCCGAGGTACTGCT
sufC F sufC R	slr1736	429 bps	AATACGGCTTCGTCACGGAT GCACGGCTTCGGAGATTGAG
sufD F sufD R	slr1737	457 bps	CTGTGCTGGAGCAAGTCAAC ACGCGAATCACGGCTGCTTG
sufS F sufS R	slr1738	422 bps	TACGCTAGGTTGCTGCAATC GCCGTTGGAAGAGGTGTTGG

2.4 Growth conditions and media

2.4.1 Growth conditions of *S. elongatus* PCC 7942

S. elongatus PCC 7942 cells were grown in modified BG11 medium according to Rippka (Rippka 1988; Stephan et al. 2000). For standard growth conditions, cultures were grown in gas wash bottles containing 250 mL medium in a stream of 2% (v/v) CO_2 in air (20 L min^{-1}). After inoculation at an optical density of 0.3 and 0.4 at 750 nm, growth was carried out for 24-144 h. Cultures were incubated in an aquarium set at 30°C under constant illumination with fluorescent light tubes Sylvana Luxline ES, 18 Watt providing a light intensity of 150 µE/m^2 x s. Growth of selected mutants was performed as described above, except that the growth medium contained the appropriate antibiotics.

2.4.2 Growth of *S. elongatus* PCC 7942 under stress conditions

For growth under iron-deficient conditions, cells grown two times for 48 h in regular BG11 medium were harvested by centrifugation for 30 min at 4,000 rpm and washed once with BG11 medium from which iron was omitted. Subsequently, cells were transferred under aseptic conditions to the same medium at an optical density of 0.3 and 0.4 at 750 nm. The iron-deficient cultures were spun down after 48, 72, and 96 h of iron-deficient growth and were harvested as described above. The iron-deficient BG11 medium was prepared generally by deleting Fe(III)-citrate.

Oxidative stress conditions were achieved by addition of 2, 5, 10 or 20 mM H_2O_2 respectively. *S. elongatus* PCC 7942 cultures were grown 24 h in BG11 medium (time 0) and then exposed to the different concentrations of H_2O_2 to induce oxidative stress for the periods of 6 and 24 h. The samples were used for RNA isolation, 77K fluorescence and photosynthetic activity determination.

2.4.3 Growth of *Escherichia coli*

Growth of *Escherichia coli* (subsequently referred as *E. coli*) was performed in liquid LBG medium (10 g/L Tryptone/Peptone from Casein, 5 g/L Yeast Extract, 10 g/L NaCl, 1g/L Glucose) cultures as well as on LBG agar-plates at 37°C. *E. coli* mutants were cultivated in LBG medium containing the appropriate antibiotics. For LBG-agar 1.5% (w/v) Agar-Agar was added.

2.4.4 Antibiotics used with *S. elongatus* PCC 7942 and *E. coli*

For the growth of *E. coli* mutants 100 µg/mL ampicillin (Ap^{100}), 50 µg/mL kanamycin (Km^{50}), 25 µg/mL chloramphenicol (Cm^{25}) and 50 µg/mL spectinomycin (Sp^{50}) were added to the growth medium.

For the growth of the *S. elongatus* PCC 7942 mutants 20 or 80 µg spectinomycin (Sp^{20}, Sp^{80}), 25 µg chloramphenicol (Cm^{25}) and 25 µg kanamycin (Km^{25}) were used.

(B) Molecular biology methods

2.5 Transformation of *E. coli*

2.5.1 Electroporation of *E. coli*

Electrocompetent *E. coli* cells were prepared according to Sharma and Schimke (Sharma and Schimke 1996). Electrocompetent cells were shock frozen in liquid nitrogen and stored at -80 °C as 100 µL aliquots.

For each electroporation 1-10 µg of plasmid DNA or 1-3 µL of a ligation reaction were mixed with 100 µL competent cells and placed in an electroporation cuvette. Electroporation was implemented with an *EasyjecT* (Equibio Ltd., Thermo Fisher Scientific, Waltham, USA) at 2.5 kV, 25 µF, and 400 Ω. Cells were mixed with 1 mL SOC medium (20 g/L Tryptone/Peptone from Casein, 5 g/L Yeast Extract, 10 mM NaCl, 10 mM $MgCl_2$, 10 mM $MgSO_4$, 2.5 mM KCl, 20 mM Glucose) and incubated at 37 °C for 45 min afterwards. Different amounts of the electroporation reaction were placed on LB-plates (see above) containing the appropriate antibiotics for selection.

2.5.2 Transformation of *E. coli*

Competent cells for transformation were prepared by $CaCl_2$ treatment (Sambrook et al. 1989). After shock freezing in liquid nitrogen, 100 µL aliquots of competent cells were stored at -80 °C.

For each transformation 1-10 µg of plasmid DNA or 5 µL of a ligation reaction were added to 100 µL freshly thawed cells. The suspension was carefully mixed and incubated on ice for 20 min. Afterwards, cells were heat-shocked at 42 °C for 1 min. Subsequently, 900 µL SOC medium (see above) was added, and the transformation was incubated at 37 °C for one hour. 100 µL of the transformation approach was placed on LBG-plates containing the appropriate antibiotics for selection.

2.6 Transformation of *S. elongatus* PCC 7942

In contrast to *E. coli* *S. elongatus* PCC 7942 features a natural DNA uptake mechanism. For transformation of *S. elongatus* WT 10 mL cell culture (O.D.$_{750 nm}$ at least 2.0) were spun down for 10 min at 2,900 x g and 25 °C. The cell pellet was washed in 10 mL BG11 medium, followed by a wash step with 10 ml 10 mM NaCl. Afterwards, the washed cells were resuspended in 1 mL BG11 medium. 0.25-2 µg of plasmid DNA were added to 500 µL cell suspension. The mixture was incubated at 30 °C in the dark over night. Subsequently, different amounts of the transformation reaction were placed on BG11-plates containing the appropriate antibiotics for selection. The plates were incubated in constant light at room temperature. After 3 weeks the first mutated cell colonies could be isolated.

2.7 Isolation of nucleic acids

2.7.1 Isolation of plasmid DNA from *E. coli* utilizing the Plasmid Miniprep Kit II™

Isolation of plasmid DNA from *E. coli* was performed with the peqGOLD Plasmid Miniprep Kit II™ from PeqLab according to the manual instructions. For further information see supplementary files.

2.7.2 Isolation of plasmid DNA from *E. coli* with HB-lysis

For HB-lysis of *E. coli* cells, 1 mL of an overnight culture was spun down for 1 min at 13,000 rpm. Cells were dissolved in 200 µL buffer 1 (50 mM Tris-HCl pH 8.0, 10 mM EDTA). Afterwards, 200 µL buffer 2 (200 mM NaOH, 1% SDS) were added and the suspension was mixed by inverting the tube for 3-4 times. 200 µL buffer 3 (2.55 mM potassium acetic acid, pH 4.8) were applied and the tube was inverted for 3 times. Afterwards, the mixture was incubated on ice for 10 min and spun down for 20 min at 13,000 rpm. The supernatant was spun down for additional 10 min at 13,000 rpm. 400-600 µL supernatant were mixed with 700 µL isopropanol by vortexing and spun down for 30 min at 13,000 rpm. The pellet was washed with 500 µL 70% (v/v) ethanol and spun down for 5 min at 13,000 rpm. Afterwards, the pellet was dried for 30 min at 60°C. Subsequently, the pellet was diluted by adding 30 µL buffer TE (10 mM Tris-HCl pH 7.5, 1 mM EDTA pH 7.5) and incubated at 50°C for 10 min. Isolated plasmid DNA was stored at -20°C.

2.7.3 Isolation of genomic DNA from *S. elongatus* PCC 7942

Isolation of genomic DNA from *S. elongatus* PCC 7942 was accomplished according to the literature (Fiore et al. 2000). This method uses the non-ionic detergent Cethyltrimethyl-ammoniumbromide (CTAB) to extract high molecular weight genomic DNA from *S. elongatus* PCC 7942. This detergent carries a positive charge which interacts with the negative charge of DNA under high salt conditions and forms soluble complexes. In the subsequent steps a decrease in the salt concentration lower than 0.5 M causes precipitation of DNA leaving other compounds, especially polysaccharides, in solution. The detergent becomes removed during ethanol based precipitations and subsequent washing steps of DNA pellets with 80% ethanol.

2.7.4 Isolation of total RNA from *S. elongatus* PCC 7942 for Northern blot analysis

Isolation of total RNA was performed as described previously (Reddy et al. 1988). The cell cultures were grown under iron sufficient conditions as well as under iron deficient growth conditions, or after H_2O_2 treatment for different times. The cells were spun down for 10 min at 2,900 x g with crushed ice. Isolation of total RNA was achieved with a hot acidic phenol extraction procedure and using the RNeasy kit (Qiagen). The mRNA was eluted from the columns 3 times in 30 µL sterile H_2O bidest.

2.8 Basic genetic methods

2.8.1 Agarose gel electrophoresis

DNA was separated on 1% TAE-agarose gels (40 mM Tris-acetic acid, 10 mM sodium acetate, 1 mM sodium EDTA). Therefore DNA samples were mixed with 0.2 volumes of loading buffer (stock solution: 10 mL TAE, 50 mL glycerol, 0.1% bromine phenol blue) before loading. Electrophoresis was carried out applying 100 V and stained with ethidium bromide (1 mg/L) to separate visualized DNA bands.

2.8.2 DNA restriction digests

Restriction of DNA was performed according to manufacturer's recommendations. A typical restriction digest consisted of 0.2-5 µg DNA, 2 µL 10 x buffer and 1 Unit (U) restriction enzyme filled to a total volume of 20 µL with sterile H_2O bidest. Restriction assays were either inactivated by incubation for 20 min at 65-80 °C or by utilizing the PCR Purification Kit (see 2.10.2).

2.8.3 Dephosphorylation of linear DNA fragments

Linear DNA fragments were dephosphorylated using shrimp alkaline phosphatase obtained from Promega as described in the producer's manual (for further information see supplementary files).

2.8.4 Phosphorylation of linear DNA fragments

Phosphorylation of PCR products was performed using T4 polynucleotide kinase received from Promega according to manufacturer`s recommendations (see supplementary files for more details).

2.8.5 Ligation of linear DNA fragments

Linear DNA fragments were ligated using the Quick Ligation Kit from New England Biolabs. Ligation was performed as described in the manual (see supplementary files for additional information).

2.8.6 Fill in of 5`-protruding DNA ends

Filling of 5`-protruding ends was done with DNA polymerase I large (Klenow) according to manufacturer`s advice. Further information on this topic is given in the supplementary files.

2.8.7 Extraction and purification of DNA from agarose gels

DNA fragments were extracted and purified from agarose gels utilizing the QIAquick Gel Extraction Kit provided from Qiagen (Hilden, Germany): The procedure was performed according to manufacturer's recommendations (see supplementary files for further details).

2.8.8 Quantification of DNA and RNA

Concentration of DNA and RNA was measured using the Nanodrop ND-1000 spectrophotometer (Peqlab). Prior to measurement, RNA was thawed on ice and vortexed.

2.8.9 RNA integrity and quality test

To test the RNA integrity and quality, a Bioanalyzer (Agilent, Böblingen, Germany) was used. One 100 ng/μL aliquot of each RNA sample was prepared and measured according to the manufacturer's recommendations (see supplementary files for further information).

2.9 Polymerase chain reaction

2.9.1 Polymerase chain reaction using BioTherm™ DNA polymerase

Polymerase chain reaction (PCR) using BioTherm™ DNA-polymerase (Genecraft) was performed to obtain gene specific DNA probes for Southern and Northern blot analyses. Typical 100 μL assays contained up to 5 μL genomic DNA, 10 μL 10 x Biotherm™ buffer, 20 mM dNTPs, up to 60 pmol primers (see table 2.4) and 5 U BioTherm™ DNA polymerase. PCRs were performed in an Eppendorf Gradient S Thermocycler (Eppendorf, Hamburg, Germany). Amplification products were separated on 1% agarose TAE gels as described above (see 2.8.1) after purification with the QIAQuick PCR-Purification Kit (Qiagen, Hilden, Germany).

2.9.2 Polymerase chain reaction using Pwo polymerase

PCR using PeqGOLD Pwo DNA-polymerase (PeqLab) was performed according to manufacturer's recommendation (see supplementary files for further information) to obtain amplification products needed for cloning experiments. PCRs were performed in an Eppendorf Gradient S Thermocycler with primers listed in table 2.4.

2.9.3 Polymerase chain reaction using Phusion™ Hot Start DNA-polymerase

PCR using Phusion™ Hot Start High Fidelity DNA-polymerase (Finnlabs, NEB) was performed as described in the manual (see supplementary files) to obtain amplification products utilized in cloning experiments. PCRs were performed in an Eppendorf Gradient S Thermocycler with primers listed in table 2.4.

2.9.4 Colony PCR

Colony PCR was used to test segregation of putative *E. coli* and *S. elongatus* PCC 7942 mutants. It was prepared according to chapter 2.9.1, but cells instead of genomic DNA were used as template. Furthermore, the PCR reaction was heated at 95°C for 5 min for cell lysis before BioTherm™ DNA-polymerase (Genecraft, Munster, Germany) was added. After the amplification, samples were purified using the QIAquick PCR Purification Kit (Qiagen).

2.9.5 Amplification of Dig-dUTP-labelled probes

Dig-dUTP-labelled gene-specific probes for Southern and Northern analyses were manufactured by PCR described in chapter 2.9.1. Genomic DNA served as template to amplify the different probes. The Dig-dUTPs (Roche, Mannheim, Germany) were added in a content of 1/10 compared to the used unlabelled dNTPs. Primers used for these reactions are shown in table 2.3 and 2.5.

2.10 Purification of DNA samples

2.10.1 PCR Purification Kit

Double or single stranded DNA with a size of 100 bp to 10 kb was purified with the help of QIAquick PCR Purification Kit (Qiagen) according to manufacturer's recommendations (see supplementary files for protocol).

2.10.2 Nucleotide Removal Kit

QIAquick Nucleotide Removal Kit (Qiagen) was used to polish nucleotides or DNA samples up to 10 kb size from enzymatic reactions, e.g. dephosphorylation and DNA restriction digests, following the kit's manual (see supplementary files for further information).

2.11 Southern blot

Genomic DNA from *S. elongatus* PCC 7942 was obtained as described above (see chapter 2.7.3) and separated on a 1.5% TAE-agarose gel. Afterwards, the gel was incubated in denaturing buffer (1.5 M NaCl, 0.5 M NaOH) 2 times for 20 min. The gel was neutralized in 1 M Tris-Cl (pH 8.0), 2 M NaCl twice for 20 min. Subsequently, DNA was transferred over night onto a Hybond N$^+$ membrane (Amersham, GE Healthcare, Freiburg, Germany) in 20 x SSC (0.3 M sodium citrate pH 7.0, 3 M NaCl). DNA was crosslinked to the membrane by UV-radiation (2.4 x 10^6 J x cm^{-1}) in a Stratalinker (Stratagene, La Jolla, CA, USA). Hybridization was performed over night in 5 x SSC, 1% (w/v) blocking reagent (Roche, Mannheim, Germany), 0.1% (w/v) sodium lauryl sarcosine, 0.02% (w/v) sodium dodecyl sulfate with Dig-dUTP-labelled probes at 68°C. The membrane was washed twice for 10 min with 2 x SSC, 0.1% (w/v) sodium dodecyl sulfate and twice for 10 min with 0.1 x SSC, 0.1% (w/v) sodium dodecyl sulfate at 68°C. Afterwards, the membrane was incubated with buffer 1 (100 mM Tris-Cl pH 7.5, 150 mM NaCl) for 1 min and transferred into buffer 2 (1% (w/v) Blocking reagent in buffer 1) for additional 20 min. Subsequently, the membrane was washed with buffer 1 for 1 min before incubation with antibody solution (4 µL Anti-digoxigenin F$_{AB}$-fragments (Roche, Mannheim, Germany) in 20 mL buffer 1) at room temperature for 30 min. The membrane was washed twice for 15 min each with buffer 1 and transferred into detection solution (45 µL nitroblue tetrazolium (75 mg/mL in 70% dimethylformamide), 35 µL 5-Bromo-4-chloro-3-indolylphosphate (50 mg/mL in dimethylformamide), 10 mL buffer 3 (100 mM Tris-Cl pH 9.5, 100 mM NaCl, 50 mM MgCl$_2$)). Staining started immediately and became visible after a few minutes. It was finished after 20 min up to 10 h. Staining was stopped by washing the membrane in buffer 4 (100 mM Tris-Cl pH 8.0, 1 mM EDTA).

2.12 Purification of RNA samples

RNA samples were purified by LiCl$_2$-precipitation. One quarter of the sample volume 4 M LiCl$_2$ and three quarter ethanol (-20°C) were added to each sample and the mixture was incubated at -70°C for 30 min. Subsequently, the mixture was spun down for 15 min at 13,000 rpm (Labofuge nano, Heraeus Sepatech). After removing the supernatant, 1 mL 70% (v/v) Ethanol (-20°C) was added and the mixture was spun down for additional 5 min at 13,000 rpm. After removing the supernatant, the pellet was dried in a SPD SpeedVac

(Thermo Savant, Thermo Fisher Scientific, Waltham, USA) for 5 min and solubilised in an appropriate volume of buffer TE.

2.13 Hybridisation of RNA

2.13.1 Northern blot

All instruments used for Northern blotting were autoclaved twice. All non heat-stable instruments were rinsed with ethanol and double autoclaved water. 10 µg RNA in a 9 µL volume were mixed with 2.5 µL 10 x MOPS (0.01 M sodium EDTA, 0.05 M sodium acetic acid, 0.2 M MOPS, pH 7.0), 12.5 µL formamide, 4 µL formaldehyde (37%), 0.5 µL ethidium bromide (10 mg/mL in water) and 1.5 µL stop solution (95% (v/v) formamide, 20 mM sodium acetic acid pH 7.0, 0.1% (w/v) bromphenol blue, 0.1% (w/v) xylene cyanol) and denatured at 68°C for ten min. Afterwards, RNA was separated in a 1.3% formalin gel (1.3 g agarose, 73 mL water, 10 mL 10 x MOPS, 17 mL 37% (v/v) formaldehyde) for up to 3 h at 80 mA. RNA was transferred onto a Hybond N^+ membrane (GE Healthcare, Freiburg, Germany) over night in 20 x SSC. RNA was crosslinked to the membrane by UV-radiation (2.4 x 10^6 J x cm^{-1}) in a Stratalinker (Stratagene, La Jolla, CA, USA).

2.13.2 Slot blot

Slot blot of RNA was performed with a Bio-Dot™ SF Microfiltration Apparatus (BioRad, Munich, Germany). 2 µg RNA were mixed with 20 µL formamide, 7 µL 37% (v/v) formaldehyde and 2 µL 2 x SSC (see chapter 2.11) and subsequently denatured for 15 min at 68°C. Samples were cooled down on ice and mixed with 80 µL 20 x SSC. Sample wells were washed twice with 10 x SSC before samples were applied as described in the manual belonging to the apparatus (see supplementary files for more details). RNA was filtered through a Hybond N^+ membrane (Amersham). Afterwards, sample wells were washed twice with 1 mL 10 x SSC. After the second rinse, the vacuum was applied for additional 5 min to dry the membrane. After removing the membrane from the Bio-Dot™ SF Microfiltration Apparatus, the membrane was dried completely at room temperature. RNA was crosslinked to the membrane by UV-radiation (2.4 x 10^6 J x cm^{-1}) in a Stratalinker (Stratagene, La Jolla, CA, USA).

2.13.3 Hybridisation of RNA with Dig-dUTP-labelled probes

Membranes were incubated for at least 2 h in CHURCH-buffer (7% (w/v) sodium dodecyl sulfate, 50% (v/v) formamide, 5 x SSC pH 7.0, 2% (w/v) Blocking reagent (Roche), 50 mM sodium phosphate buffer pH 7.0, 0.1% (w/v) lauryl sarcosine) at 52°C. 3 µg Dig-dUTP-labelled DNA probe were diluted in 7 mL CHURCH-buffer and denatured at 90°C for 10 min and subsequently applied to the membranes. Hybridisation was performed at 52°C over night. Membranes were washed for 5 min in 2 x SSC and 0.1% (w/v) sodium dodecyl sulfate at room temperature. Afterwards, membranes were washed first in 0.1 x SSC and 0.1% (w/v) sodium dodecyl sulfate at 52°C for 20 min. After changing the buffer the membranes were washed for 20 min at 68°C.

2.13.4 Colorimetric detection

After hybridisation and stringent washing steps, membranes were incubated with buffer 1 (see chapter 2.11) for 1 min at room temperature, before incubation with antibody solution (4 µL Anti-Digoxigenin F_{AB}-fragments (Roche, Mannheim, Germany) in 20 mL buffer 1) for 30 min at room temperature. Membranes were washed twice with buffer 1 for 15 min each. Subsequently, the membranes were transferred into detection solution (see chapter 2.11). Staining started after a few min and finished after up to 10 h. Staining was stopped by washing the membrane in buffer 4 (see chapter 2.11).

2.13.5 Detection with CDP-Star™

Detection with CDP-Star™ (Roche) was performed according to manufacturer's recommendations (see supplementary files for manual). The blots were developed by exposure to FUJI SuperRX X-Ray films (Hartenstein, Würzburg, Germany).

2.14 DNA-microarray experiments

2.14.1 Isolation of total RNA from *S. elongatus* PCC 7942 for DNA-microarray experiments

Total RNA was isolated from cell pellets harvested as described above. Frozen cells were resuspended in 200 µL Tris-HCl pH 8.0, 700 µL RLT-buffer, provided by the RNeasy Mini Kit (Qiagen) and 7 µL β-mercaptoethanol. The cell suspension was transferred to Fast Protein Tubes (Lysing Matrix B, Q Biogene, Carlsbad, CA, USA), and cells were disrupted using the ribolyzer (Hybaid, Heidelberg, Germany) (30 s, level 6.5). Total RNA was isolated using the RNeasy Mini Kit (Qiagen) and treated with DNase (RNase-free DNase Set, Qiagen). The isolated RNA was concentrated using Microcon-30 filters (Millipore, Eschborn, Germany) and the RNA concentration was measured with a Nanodrop (ND-1000 Spectrophotometer, PeqLab).

2.14.2 Target labelling

Preceding the labelling of the targets, total RNA was transcribed into cDNA by a reverse transcriptase assay. For primer annealing 10 µg of total RNA and 10 µg random hexamer primers (Qiagen-Operon, Hilden, Germany) were incubated 10 min at 70°C and subsequently 5 min on ice. For reverse transcription, the assay, which consisted of the RNA-hexamer assemblies, 1 x RT first strand buffer, 10 mM DTT, RNase inhibitor (20U), Superscript III RT (300U, Stratagene), 0.06 x dNTP stock solution including aa-dUTP (4:1 aa-dUTP/dTTP nucleotide mix) (dNTPs: PeqLab, aa-dUTP: Sigma-Aldrich, Taufkirchen, Germany), was incubated for 2 h at 42°C. The aminoallyl-labelled cDNA was coupled with *N*-hydroxysuccinimidyl ester dyes (Cy3- and Cy5-NHS esters, Amersham Biosciences, Freiburg, Germany) and purified by CyScribe GFX Purification Kit (GE Healthcare, Freiburg, Germany). The labelled cDNAs were kept in brown Eppendorf tubes to protect the dyes from light. The achievement of labelled cDNA was controlled by Nanodrop measurements. The labelling of aminoallyl-modified cDNA was performed according to DeRisi et al. (DeRisi et al. 1997) and resulted in 100 µL of labelled cDNA, respectively.

2.14.3 Hybridisation and wash protocols

Prior to hybridisation, the Nexterion E Slides were processed to block free epoxy groups. After warming to RT for 5 min the following washing steps were performed: 5 min at RT in rinsing solution 1 (250 µL Triton X100 were dissolved in 250 mL MilliQ H_2O at 80°C for 5 min and cooled down to RT), 2 min at RT in rinsing solution 2 (50 µL 32% HCL were mixed with 500 mL MilliQ H_2O) twice, 10 min at RT in rinsing solution 3 (25 mL 1 M KCL were mixed with 225 mL MilliQ H_2O), and finally 1 min at RT in MilliQ H_2O. Subsequently, the slides were incubated in a glass container at 50°C for 15 min in pre-warmed Nexterion Block E blocking solution (Schott) (50 mL 4 x blocking solution and 47 µL 32% HCL were mixed with 150 mL MilliQ H_2O) and swayed for at least 5 min. Thereafter, the slides were washed for 1 min at RT in MilliQ H_2O and than immediately dried in a microplate centrifuge (Heraeus Sepatech) at 1,200 rpm for 3 min.

Before hybridization, the total amount of 100 µL of every labelled cDNA target were each evaporated to a volume of about 10 µL and refilled to a volume of 100 µL with Dig Easy Hyb hybridization solution (Roche, Mannheim, Germany), containing 1.5 µL (5 µg/µL) sonicated salmon sperm DNA (Amersham Biosciences, Piscataway, NJ, USA). The hybridization samples were incubated at 65°C for 5 min. The hybridization was performed by using a HS4800 hybridization machine. Subsequently, the following washing steps were performed: The slides were transferred to pre-warmed 2 x SSC, 0.2% (w/v) SDS (42°C) in a black plastic rack and were swayed for 1 min. Thereafter, the slides were washed twice in 0.2 x SSC and 0.1% (w/v) SDS (RT) in a black plastic rack und were swayed for 1 min. Then the slides were washed twice in 0.2 x SSC (RT) in a black plastic rack and were swayed for 1 min. Finally the slides were transferred to 0.05 x SSC (21°C) in a black plastic rack and were swayed for 1 min. Subsequently, the slides were dried immediately in a microplate centrifuge (Heraeus) at 1,200 rpm for 3-5 min. The samples were always kept in the dark and were not exposed to high ozone concentrations to avoid bleaching of the Cy5 channel.

2.14.4 Image acquisition and data analysis

The scanning of the Cy3 (A555) / Cy5 (A647)-hybridized microarrays was performed with a LS Reloaded Microarray Scanner (Tecan Deutschland GmbH, Crailsheim, Germany). The acquisition of images and analysis of the data were performed as described in Brune et al. (Brune et al. 2006). For spot detection, image segmentation and signal quantification, meaning determination of the mean signal and the mean local background intensities for each spot of the microarray, were assigned by using the *ImaGene 6.0* software (Bio Discovery, Los Angeles, CA, USA.). After subtraction of the local background intensities from the signal intensities, the log2 value of the ration of intensities was calculated for each spot according to the formula $M_i=\log_2(R_i/G_i)$. In particular, $R_i=I_{ch1i}-Bg_{ch1i}$ and $G_i=I_{ch2i}-Bg_{ch2i}$, where I_{ch1i} or I_{ch2i} is the intensity of a spot in channel 1 or channel 2 and Bg_{ch1i} or Bg_{ch2i} is the background intensity of a spot in channel 1 or channel 2, respectively. In both channels the average intensity was calculated for each spot according to the formula $A_i=\log_2\sqrt{(R_iG_i)}$ (Dudoit et al. 2002). Thereafter, a normalization method based on robust local regression, and accounts for intensity and spatial dependence in dye biases, was applied according to Yang et al. (Yang et al. 2002). Differentially expressed genes were identified by t-statistics

(Dudoit et al. 2002). Normalization of the data and t-statistics were carried out by using the *EMMA 2.2* microarray data analysis platform (Dondrup et al. 2003) (http://www.cebitec.uni-bielefeld.de/groups/brf/software/emma_info/). Genes were regarded as differentially expressed if $p \leq 0.051$ and $M \geq 0.98$ or $M \leq -0.98$. Each experiment was performed with three biological replicates including one dye-swap.

(C) Biochemical methods

2.15 Determination of cell growth

Cyanobacterial growth rates were determined by measuring the optical density at 750 nm ($OD_{750\,nm}$) of cell cultures in plastic cuvets (1 cm slit) using a spectrophotometer (UV 1202 UV-VIS, Shimadzu). Samples were measured against BG11 medium. An $OD_{750\,nm}$ of a cell culture corresponded to a cell density of about 0.1 µL packed cell volume per mL (Flores et al. 1982).

2.16 Determination of pigment content

Chlorophyll was determined in a methanol extract according to Grimme and Boardman (Grimme and Boardman 1972). The Determination of phycocyanin, allophycocyanin, and phycoerythrin was performed in a cell-free extract after addition of streptomycinsulfate according to Tandeau de Marsac and Houmard (Tandeau de Marsac and Houmard 1988). The color of the cell cultures was documented by scanning 270 µL culture in a microtiterplate (300 µL) using a hp Scanjet 7400c (Hewlett Packard, Paolo Alto, California).

2.17 Determination of protein content

The protein content of samples was determined using the method according to Smith et al. (Smith et al. 1985). A calibration curve of 5 to 40 µg BSA was used to calculate the protein content of given samples. The protein determination described by Bradford (Bradford 1976) was used if the Smith test showed interference with certain buffer substances or metabolites in the samples of interest, e.g. thylakoid membranes isolated by using a sucrose gradient (see chapter 2.21.2).

2.18 Preparation of cell suspensions

For preparation of cell suspensions, cells were harvested by centrifugation for 30 min at 2,200 x g at room temperature and were washed once or not. Subsequently, the cells were resuspended in the medium of choice, as e.g. BG11 as used for growth or in buffer as 0.1 M sodium phosphate buffer, pH 7, or 0.1 M HEPES-NaOH, pH 7, to give a cell density of 100 µl packed cell volume (PCV) per mL. Such cell suspensions were suited for determining the photosynthetic O_2 evolving activity of whole cells with sodium bicarbonate as electron acceptor or for the preparation of cell-free extracts.

2.19 Preparation of cell-free extracts

2.19.1 Preparation of cell-free extracts of S. elongatus PCC 7942 with a French Press

For preparation of cell-free extracts, cells were harvested by centrifugation at room temperature, washed once with the appropriate buffer, e.g. 10 mM sodium phosphate buffer, pH 7.0, and then resuspended in the same buffer to give a PVC of 100 µL cells mL^{-1}. The cell suspension was then passed twice through a pre-chilled French Pressure Cell (SLM Aminco, Urbana, IL, USA.) at 138 MPa (20,000 Psi). Unbroken cells were removed by centrifugation at 4,000 x g for 5 min at 4 °C. The supernatant solution was used for immuno-blotting. For thylakoid preparations HEPES and Mes buffers were used.

2.19.2 Preparation of cell-free extracts of S. elongatus PCC 7942 using a Ribolyzer

Pelleted cells were resuspended in 20 mM Mes pH 6.5, 10 mM $MgCl_2$, 10 mM $CaCl_2$ and 0.5 M D-mannit and filled in a 1.5 mL screw cap tube. Equal amounts of glass beads (Ø 0.02 mm) were added. Cell lysis was performed for 5 times 15 s at 6,500 ms^{-1} with cooling on ice after each cycle in a Hybaid Ribolyser (Thermo Fisher Scientific, Waltham, USA.). After centrifugation for 1 min at 13,000 x g in a table top centrifuge (Heraeus Sepatech) the protein containing supernatant was transferred into a fresh tube and stored at -20 °C for later usage.

2.19.3 Preparation of cell-free extracts of E. coli

E. coli cells were harvested by centrifugation at 4°C with 10,000 x g for 10 min and resuspended in appropriate buffer (20 mM HEPES pH 7.0 or the corresponding lysis buffer if the samples were used for protein purification via Ni-NTA affinity chromatography (see chapter 2.25.2) to give a cell density according to a PCV of 100 µL cells mL^{-1}. The cell suspension was then passed twice through a pre-chilled French Pressure Cell (SLM Aminco, Urbana, IL, USA.) at 138 MPa (20,000 Psi). Unbroken cells were removed by centrifugation at 4,000 x g for 5 min at 4 °C. The cell-free extract was stored at 4°C.

2.20 Isolation of subcellular fractions

2.20.1 Isolation of thylakoid membranes

For the isolation of the membrane associated proteins, cell-free French press extracts were prepared followed by a centrifugation step at 4°C and 28,000 x g for 20 min. The pellet contained all insoluble cell parts, mainly the thylakoid membranes, whereas the supernatant contained the soluble proteins. If the samples were used performing immuno blots, 10 mM sodium phosphate buffer were utilised. For preparation of PS I complexes, 20 mM Mes, pH 6.5, 10 mM $CaCl_2$, 10 mM $MgCl_2$, 0.5 D-mannit were used. If the thylakoid membranes should be used for Blue Native gelelectrophoresis, cells were diluted in 50 mM HEPES-NaOH pH 7.0, 5 mM $MgCl_2$, 25 mM $CaCl_2$ and 10% (v/v) glycerol.

2.20.2 Isolation of periplasmic proteins

The isolation of periplasmic proteins was carried out according to the procedure given by Fulda et al. (Fulda et al. 2000; Fulda et al. 1999). The procedure is based on a treatment of S. *elongatus* PCC 7942 cells with a cold osmotic shock. The resulting periplasmic fraction was concentrated using Microcon-10 filters (Millipore, Eschborn, Germany). The corresponding pellet, containing the spheroplasts, was resuspended in 10 mM sodium phosphate buffer, pH 7.0.

Alternatively, the periplasmic proteins were isolated according to Block and Grossman (Block and Grossman 1988). This method is also based on cell lysis applying a cold osmotic. However, the periplasmic proteins were released in sucrose-containing 30 mM Tris-Cl buffer pH 7.5 according to higher protein stability.

2.20.3 Isolation of the outer membrane, the cytoplasmic membrane, and the thylakoid membrane

The different membrane systems of S. *elongatus* PCC 7942 were isolated according to the procedure described by Omata and Murata (Omata and Murata 1984). This procedure is based on a separation of the membranes by their different specific density. The membrane systems were separated in a sucrose gradient by centrifugation using an ultra centrifuge (Beckman Coulter, Palo Alto, California) at 4°C for 18 h at 130,000 x g in a swinging bucket rotor TST 28.38 (Kontron, Zurich, Switzerland). Separating French press extracts, distinct fractions for the cytoplasmic membranes, thylakoid membranes, outer membranes and the soluble proteins containing the phycobilisomes were obtained.

2.21 Isolation of PS I complexes from *S. elongatus* PCC 7942

Isolation of membrane complexes can be achieved with non-ionic detergents. Such detergents can form protein-detergent micelles, in which the detergent in part replaces the lipids of the membrane. The detergent surrounds the complex in such a way that the hydrophobic part of the detergent is directed to the inside of the micelle, while the hydrophilic part is located at the outside and thus, making the membrane complex soluble.

PS I complexes from *S. elongatus* PCC 7942 were isolated as described before (El-Mohsnawy 2007) or with some modifications as given in the literature (Wenk and Kruuip 2000). The work was done in the lab of Prof. Dr. Mathias Rögner at the University of Bochum. The chromatography was performed with a high performance liquid chromatography (HPLC) BioCAD 700 E (Applied Biosystems, Foster City, California). The procedure contains two chromatography steps: hydrophobic interaction chromatography (HIC) of PS I-enriched thylakoid membrane fractions, obtained via solubilization with 0.5% (w/v) of β-dodecyl maltoside and 0.2% $(NH_4)_2SO_4$, led to the separation of PS II particles and two different PS I particle species: PS I monomers and PS Itrimers. Subsequently, the PS I trimer- containing fraction was separated via anionic exchange chromatography (IEC). The HIC was performed using a POROS 50OH column. For the IEC a UNO Q II column (both columns manufactured by Bio PerSeptive, Wiesbaden, Germany) was used. The resulting PS I trimers were concentrated and desalted by centrifugation using Centricon-30 filters (Millipore, Eschborn, Germany). Figure 2.2 gives an overview of the whole procedure.

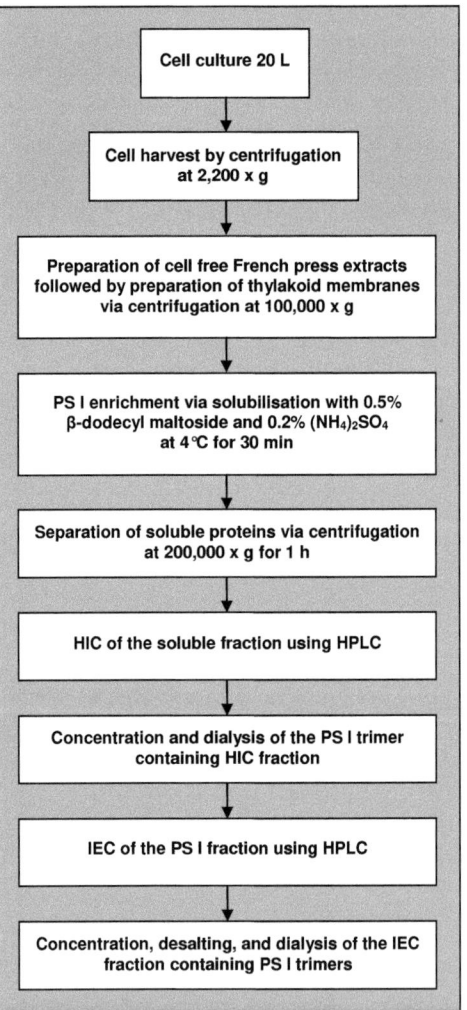

Figure 2.2: PS I isolation via HPLC.

2.22 Determination of photosynthetic activity

2.22.1 Determination of photosynthetic oxygen evolution

Determination of the total photosynthetic activity in intact cells was done in a Clark-type electrode (Rank Brothers, Bramsham, U.K.) at a polarization voltage of 600 mV. The measurements were performed according to previous protocols (Stephan et al. 2000). Cells grown for 24, 48, 72 and 96 h were utilized.

2.22.2 Determination of respiratory oxygen uptake

Determination of the respiratory oxygen uptake in intact cells was done in a Clark-type electrode (Rank Brothers, Bramsham, U.K.) at a polarization voltage of 600 mV. The measurements were performed according to previous protocols (Stephan et al. 2000). Cells grown for 24, 72 and 120 h were utilized.

2.22.3 Determination of the activity of PS I

P_{700} rereduction was measured at 820 nm with a pulse-amplitude-modulated system equipped with an ED-800T emitter-detector unit (PAM101-103, Walz, Germany) as described (Eisenhut et al. 2007). The measurements were done at the University of Rostock in cooperation with Prof. Dr. Martin Hagemann and Prof. Dr. Hendrik Schubert. Far-red radiation was provided by means of a cut on/cut off filter system with transmission between 705 and 735 nm only. Dark-adapted cells were treated with increasing far-red intensities, at each intensity P700 oxidation state was measured as relative absorbance equilibrium signal (Schreiber et al. 1988). After reaching equilibrium, a strong (approximately 12,000 mmol photons m^{-2} s^{-1}) white light pulse (5 s) was applied to activate PS II, causing a drop in the P_{700} absorbance signal. The P_{700} absorbance signal in presence of white (PS II) light was used for analysis by averaging the readings between 1 and 4 s (240 values) after onset of the pulse. Far-red intensity was measured by means of a spectroradiometer (SR9910, Macam Photometrics Ltd.).

2.22.4 Determination of the effective quantum yield of PS II

To measure the effective quantum yield of PS II, a portable chlorophyll fluorometer (Mini PAM, Heinz Walz GmbH, Effeltrich, Germany) was utilized. 250 mL cells were centrifuged for 30 min in a Stock centrifuge (see above). Cell pellets were resuspended in BG11 medium and adjusted to give a cell density of 100 µL packed cells per mL. 200 µL of this cell suspension were applied to a microtiter plate and placed on a glass plate. Illumination (100 µmol photons m^{-2} s^{-1}) was performed from the bottom up for 5 min. After starting the measurement, the fluorescence yield (F) and the maximal yield (Fm) were measured from above by a microfiber optic and the according photosynthesis yield (Y = ΔF/Fm) was calculated.

2.22.5 77K pigment fluorescence measurements

Measurements of 77K low temperature pigment fluorescence allows for estimation of the content of PS II in relation to PS I. 77K measurements also allows for detection of phycobilisome amount, integrity and localization.

After growth for 48, 72, and 96 h under iron sufficient and deficient conditions, cell cultures were diluted in fresh BG11 to give an equal chlorophyll a content within a range of 5-10 µg Chl mL^{-1}. This cell suspension was then transferred to cryotubes and frozen immediately in liquid nitrogen. 77K chlorophyll a fluorescence emission spectra were recorded in a PerkinElmer LS 50B luminescence spectrometer (PerkinElmer, Waltham, Massachusetts) with an excitation wavelength of 435 nm (slit 10 nm) within an emission range from 600-750 nm (slit 5 nm, 515 nm cut-off filter) at 200 nm/min. Emission spectra were also detected after excitation via phycobilisome antenna with a wavelength of 600 nm. The emission spectra contain typical peaks at ~655 nm (phycobilisomes), 685 nm (phycobilisomes and CP43), 695 nm (P_{680} and CP47) and ~715 nm (P_{700}).

2.23 Recombinant expression of IdiC

2.23.1 Heterologous expression of IdiC in *E. coli*

For heterologous expression of IdiC in *E. coli* two different expression systems were used. At first the entire PCR-derived *idiC* sequence was cloned into a pQE-82L overexpression vector (Qiagen). Due to the vector sequence, a 6-fold histidine tag (6 x His-tag) was attached to the N-terminal end of the protein. All attempts to express the tagged protein in *E. coli* strains DH10B and M15 resulted in formation of protein aggregates. Thus, the protein was solubilised with 0.3% (w/v) SDS and subsequently purified using Ni-NTA affinity chromatography (see chapter 2.25). The purified protein was used to raise an antibody in rabbit (see chapter 2.26).

As a second attempt, the *his6idiC* sequence was cut out of pQE82-*idiC* and inserted in the overexpression vector pET32a(+) (Novagen). The pET-vector system provides a thioredoxin tag (Trx-Tag) at the C-terminal end of the heterologously expressed protein to enhance its solubility. After transformation in *E. coli* strain BL21(DE3) (Novagen), the expression of the recombinant protein was performed in LBG medium containing 100 µg/mL ampicillin and 10 µM FeCitrate at 18°C for 19 h, induced by adding IPTG to a final concentration of 0.1 mM and 0.2 mM. Cells were harvested and French pressed described in chapter 2.20.3. The pelleted protein was used for iron determinations with ICP and EPR (see chapter 2.37/2.38).

2.23.2 Homologous expression of IdiC in *S. elongatus* PCC 7942

For homologous recombinant expression of IdiC in *S. elongatus* PCC 7942 the vector p39 was used. This vector represents a fusion product of the plasmids pAM2255 and pAM2314 (Texas AM strain file, Susan Golden, Texas A&M University) generated by insertion of a 2214 kb *Bgl*II fragment from pAM2255 in the *Bam*HI site of pAM2314. The corresponding p39 vector contains *Ptrc* and *lacI* sequences of pAM 2255 as well as the neutral site 1 (NS1) of pAM 2314. The entire *his6idiC* sequence was amplified via PCR and subsequently cloned in the single *Nco*I site of p39. After transformation in *S. elongatus* PCC 7942 WT, the vector sequence including *Ptrc*, *his6idiC*, and *laqI* was inserted in the genome via homologous recombination using the ambient NS1 sequence. The obtained *S. elongatus* mutants were cultivated in BG11 medium containing 20 µg/mL Spectinomycin for 72 h. After addition of additional Fe(III)Citrate to a final concentration of 30 µM, the cultures were incubated for 1 h under standard conditions (see chapter 2.4.2). Subsequently, the recombinant expression was initiated by adding 2 mM IPTG. The cultures were harvested after 3 h after induction of IdiC expression.

2.24 Protein purification

2.24.1 Solubilisation of recombinant IdiC protein

Inclusion bodies of recombinant expressed IdiC protein were solubilised using sodium dodecyl sulfate (SDS) or β-dodecyl maltoside (β -DM). If SDS was used, it was added to the protein solution to give a final concentration of 0.3% (w/v). The solution was mixed using a potter and stirred gently for 1 h at room temperature. Subsequently, the soluble proteins were separated by a centrifugation step at 10,000x g for 30 min.

If β-DM was used, it was added to a final concentration of 2% (w/v). The solution was also mixed using a potter and stirred gently for 1 h at 4°C. This step was repeated for one time. After centrifugation at 10,000x g and 4°C for 30 min the supernatant contained the solubilised proteins.

2.24.2 Batch purification of 6 x His-tagged IdiC using Ni-NTA affinity chromatography

Recombinant expressed His-tagged IdiC protein was purified by Ni-NTA affinity chromatography using gravity flow columns provided by Qiagen. The purification was performed under native conditions at 4°C as well as under denaturing conditions at room temperature following manufacturer's recommendations (see supplementary files).

2.24.3 Preparative SDS PAGE and electroelution

Up to 1.5 mL protein containing sample were mixed with the same volume denaturing buffer (5.4 mL H_2O, 0.6 mL 1 M Tris-HCl pH 6.8, 2 mL 10% (w/v) SDS, 1 mL glycerol, 0.5 mL 0.5% (w/v) brome phenol blue, 0.5 mL β-mercaptoethanol) and incubated for 20 min at 60°C. Samples were separated on 12% tris glycine SDS PAGE (see chapter 2.28.1). After separation, the gel was stained with 0.3 M $CuCl_2$ for 5 min. Afterwards, the band of interest was excised from the gel, and destained 3 times 5 min in H_2O by gently shaking. Gel pieces were hackled and electroelution was performed with an Elutrap™ Electroelution System (Schleicher & Schüll, Dassel, Germany) following the protocol belonging to the system (see supplementary files for whole protocol).

2.24.4 Dialysis

Dialysis was primarily used to reduce the detergent concentration of protein solutions. At first the dialysis tubing (Medicell International LTD., London, England, MWCO: 12-14 kDa) was incubated in H_2O bidest for 10 min. Afterwards, the sample was filled in the tubing and both ends were sealed with dialysis clips. The dialysis tubing was applied to at least a 100-fold of sample volume dialysis buffer. The sample was gently agitated at room temperature or at 4°C, while the detergent or salt concentration of the buffer was reduced step by step.

2.25 Generation of a polyclonal antibody against IdiC

The production of a polyclonal antibody against the IdiC protein of *S. elongatus* PCC 7942 was performed by Pineda Antikörper-Service (Berlin, Germany). Heterologously expressed IdiC was purified by Ni-NTA affinity chromatography and preparative SDS-PAGE and subsequently used to immunize four rabbits. Bleedings were taken at 61, 91, and 120 days. See supplementary files for immunisation protocol.

2.26 Production of peptide antibodies against NdhA and NdhB

The peptide antibodies against NdhA and NdhB of *S. elongatus* PCC 7942 were produced by Pineda Antikörper-Service (Berlin, Germany).

Table 2.6: Amino acid sequences of NDH-peptides used for antibody generation in rabbits.

Antiserum	Peptide sequence
NdhA	CMSGYASNNKYSLLGGLR
NdhB	CLMTGYMKRDPRSNEAALKY

Two rabbits were immunized per antibody with the synthesized and crosslinked oligopeptides. Bleedings were extracted at 61, 91, and 120 days.

2.27 Generation of peptide antibodies against FNR

The anti-FNR peptide antisera anti-FNR 1 and anti-FNR 9a were raised against keyhole limpet-coupled intrinsic *S. elongatus* PCC 7942-FNR oligopeptides of 16 amino acid (aa) residues length: anti-FNR 1 antiserum against a 27-42 aa *N*-terminal CpcD-like domain of FNR; anti-FNR 9a antiserum against a 385-399 aa *C*-terminal domain.

Table 2.7: Amino acid sequences of FNR-peptides used for antibody production in rabbits.

Antiserum	Peptide sequence
FNR 1	CRQAEGEPSDSSIRRS
FNR 9a	CWSDYQRTLKKAGRWH

Here again, two rabbits were immunized per antibody with the synthesized and crosslinked oligopeptides. Bleedings were taken also at 61, 91, and 120 days.

2.28 Precipitation of proteins

2.28.1 Acetone precipitation

For acetone precipitation samples were filled up to a total volume of 100 µL with H_2O bidest. After adding 900 µL acetone (-20°C) precipitation was performed for 30 min at -20°C. Afterwards, samples were centrifuged for 10 min at 13,000 rpm at 4°C (Labofuge pico, Heraeus Sepatech). After removing the supernatant, pellets were dried in a SpeedVac (Thermo Savant, Thermo Fisher Scientific, Waltham, USA) for 15 min. Dried pellets were dissolved in selected buffer for further usage.

2.28.2 Chloroform-methanol precipitation

Chloroform and methanol were mixed 2:1 and chilled at -20°C. Samples were filled up to a total volume of 100 µL with H_2O and 900 µL chloroform-methanol were added. Precipitation was carried out for 30 min at -20°C. Subsequently, samples were centrifuged for 10 min at 4°C and 12,000 x g. The upper phase was mixed with 1 mL methanol by vortexing thoroughly. Centrifugation was repeated as described. Supernatant was removed and the pellet was dried in a SpeedVac (Thermo Savant, Thermo Fisher Scientific, Waltham, USA) for 15 min. Dried pellets were diluted in selected buffer according to their later usage.

2.28.3 Trichloroacetic acid precipitation

The samples were mixed 1:1 with 100% trichloroacetic acid and vortexed thoroughly. Precipitation was performed for 15 min at -20°C. Afterwards, the samples were centrifuged at 10,000 x g for 5 min. After the resulting protein pellet was washed with ethanol/ether (1:1), it was resuspended in selected buffer for further usage.

2.29 Sodium dodecyl sulfate gel electrophoresis (SDS PAGE)

2.29.1 Tris glycine SDS PAGE

Tris glycine SDS PAGE was performed using the Mini-PROTEAN II electrophoresis systems (BioRad, Munich, Germany) for 0.75 mm thick gels and larger self made gels (150 x 80 x 1 mm). Samples were mixed with an equal volume denaturing buffer containing 10 mM Tris-HCl, 10% (v/v) glycerol, 2% (w/v) SDS, 0.01% (w/v) brome phenol blue, 100 mM DTT (Laemmli 1970) and incubated for 30 min at 60°C. Alternatively, samples were mixed 1:1 with denaturing buffer being composed of 10 mM Tris pH 6.8, 20% (v/v) glycerol, 2% (w/v) SDS, 0.05% (w/v) brome phenol blue, 5% (v/v) β-mercaptoethanol (Helmann 1998) and incubated at 100°C for 5 min. Subsequently, samples were separated on 10% or 12% SDS-polyacrylamide gels at room temperature.

2.29.2 Urea SDS PAGE

Urea SDS PAGE was carried out as described in chapter 2.30.1, except that 2 M or 4 M urea was added to the gel and the electrophoresis running buffer.

2.29.3 Tris tricine SDS PAGE

Tris tricine SDS-PAGE was performed as described before (Schaegger and von Jagow 1987). Mini-PROTEAN II electrophoresis systems (BioRad) for 0.75 mm thick gels were used as well as larger self made gels (150 x 80 x 1 mm). The samples were incubated 1:1 in denaturing buffer (10 mM Tris-HCl pH 6.8, 25% (v/v) glycerol, 2.5% (w/v) SDS, 0.02% (w/v) Serva Blue G, 10% β-mercaptoethanol) for 30 min at 60°C.

2.30 Native PAGE

Polyacrylamide gel electrophoresis under non-denaturing conditions was arranged as described previously (Forchhammer and Tandeau de Marsac 1994). Native PAGE was performed using Mini-PROTEAN II electrophoresis systems (BioRad, Munich, Germany) for 0.75 mm thick gels.

2.31 2D Blue Native PAGE

2-dimensional Blue Native PAGE (BN-PAGE) was accomplished as specified previously (Jänsch 1996). Thylakoid membranes were prepared as described (see chapter 2.21.1) from cells grown for 120 h in regular BG11 medium and 96 h in iron depleted medium respectively. For both electrophoresis dimensions a PROTEAN II electrophoresis chamber (BioRad) for 1.5 mm thick gels was used.

The first BN-PAGE dimension was used performing the second gel dimension as well as for immuno blots. The second dimension was used for immuno blots as well. Protein spots were also cut out and subsequently detected via MALDI-TOF MS (see chapter 2.36.3).

2.32 Staining of protein gels

2.32.1 Coomassie Brilliant Blue stain

After electrophoresis, protein gels can be simultaneously fixed and stained using this method. Gels were soaked in staining solution (0.025% (w/v) Coomassie Brilliant Blue G250, 0.025% (w/v) Coomassie Brilliant Blue R250, 10% (v/v) methanol, 7% (v/v) acetic acid) with gentle shaking over night. Gels were destained by washing in 7% (v/v) acetic acid for several times.

2.32.2 Silver stain

Staining polypeptides with silver salts after separation via SDS-PAGE is based on the differential reduction of silver ions that are bound to the side chains of amino acids. This method is capable of detecting as little as 0.1-1 ng of protein in single band.

Protein gels were incubated in fixing buffer (40% (v/v) methanol, 13.5% (v/v) formaldehyde) for 10 min and washed twice with H_2O for 5 min. Subsequently, gels were deoxidized with 0.02% (w/v) sodium dithionate for 1 min followed by two washing steps with H_2O for 20 s each. Staining was carried out for 10 min in 0.1% (w/v) silver nitrate. Afterwards, gels were washed with H_2O for 10 min. Gels were transferred into developer (3% (w/v) sodium carbonate, 0.0004% (w/v) sodium dithionate, 0.05% (v/v) formaldehyde) until staining was accomplished for up to 15 min. Staining was stopped by washing with stop solution (25% (v/v) isopropanol, 10% (v/v) acetic acid).

2.33 Immunoblot analysis

2.33.1 Protein transfer

After separation using polyacrylamide gels as described above, proteins were transferred onto nitrocellulose membranes (Nitrocellulose BA85, Schleicher & Schüll, Dassel, Germany) by capillary transfer in transfer buffer (10 mM Tris-HCl pH 8.8, 2 mM EDTA, 50 mM NaCl, 0.1 mM DTT) over night at room temperature. Protein transfer could be controlled by staining with Ponceau Red™ (Serva) solution. The membranes were incubated for 30 min in staining solution (0.1% (w/v) Ponceau Red, 1.5% (v/v) trichloric acid, 1.5% (v/v) sulfosalic acid) and destained by washing with 5% (v/v) acetic acid.

2.33.2 Colorimetric detection using 4-chloro-1-naphtol

After blotting, membranes were washed 2 times for 10 min with CMF-PBS (137 mM NaCl, 2.7 mM KCl, 8 mM Na_2HPO_4 x 2 H_2O, 1.5 mM KH_2PO_4). Subsequently, the membranes were blocked with 5% (w/v) low-fat milk powder for 1 h. The membranes were washed once for 15 min and twice for 5 min with washing buffer (0.5% (w/v) low-fat milk powder, 0.05% (w/v) Tween-20 in CMF-PBS) and subsequently incubated with the specific primary antibody diluted in 0.5% (w/v) BSA in CMF-PBS for 2.5 h. Membranes were washed as described previously with washing buffer, before incubation with horse radish peroxidase (HRP) conjugated secondary antibody, diluted 1:1,000 in 0.5% (w/v) BSA in CMF-PBS, for 60 min. Afterwards, membranes were washed as described with washing buffer, one time for 5 min with CMF-PBS and one time for 5 min with 50 mM Tris-HCl pH 7.35. Subsequently, developer (12 mg 4-chloro-1-napthol diluted in 4 mL methanol, 20 mL 50 mM Tris-HCl pH 7.35, 4 µL 30% (v/v) H_2O_2) was added. Staining time was depending on the used antibody and stopped by washing the membranes in H_2O bidest.

2.33.3 Colorimetric detection using nitroblue tetrazolium and bromo-4-chloro-3-indolylphosphate

Compared to the previously described staining method, this staining is characterised by its higher sensitivity. The membranes were treated as described in chapter 2.34.2, but instead of horse radish peroxidase conjugated secondary antibody an alkaline phosphatase (AP) conjugated secondary antibody was used, diluted 1:1,000 in 0.5% (w/v) BSA in CMF-PBS. After washing the membranes twice with washing buffer, once with CMF-PBS and 100 mM Tris-Cl, pH 9.5, 100 mM NaCl, 50 mM $MgCl_2$, the detection was performed as described for Southern blotting in chapter 2.11.

2.33.4 Chemiluminescent detection

Chemiluminescent detection represents the detection method allowing for the highest sensitivity compared to the colorimetric detection methods. However, because of unclassified chemilumiscence-stabilizing the price is much higher. The membranes were handled as described in chapter 2.34.2. Horse radish peroxidase (HRP) conjugated secondary antibody was diluted 1:2,000. After the last washing step using CMF-PBS, protein detection was performed with the ECL Detection Kit (GE Healthcare, Freiburg, Germany) as described in the manual (see supplementary files) and exposed to FUJI SuperRX X-Ray films (Hartenstein, Würzburg, Germany).

2.33.5 Antibodies used in this work

The following antibodies were used in this work:

Table 2.8: Antibodies used in this work.

Antibody	Dilution	Organism	Remarks/Literature
anti-IdiA	1:2,000	S. elongatus PCC 7942	Michel, 1996
anti-IdiB	1:300	S. elongatus PCC 7942	Yousef, 2003
anti-IdiC	1:500	S. elongatus PCC 7942	this work
anti-IsiA	1:1,000	Synechocystis sp. PCC 6803	Hess, Freiburg, Germany
anti-IsiB	1:500	unknown	Mühlenhoff
anti-IrpA	1:750	S. elongatus PCC 7942	Höcker, 2008
anti-IrpB	1:1,000	S. elongatus PCC 7942	Höcker, 2008
anti-DpsA	1:2,000	S. elongatus PCC 7942	Bullerjahn, Ohio, USA
anti-D1	1:2,000	Avena sp.	Gärtner, 1998
anti-PsbO	1:3,000	Synechocystis sp. PCC 6803	Kuhlmann, 1987
anti-PsaA/B	1:1,000	Synechocystis sp. PCC 6803	J. Kruip, Bochum, Germany
anti-NdhA	1:750	S. elongatus PCC 7942	directed in this work against synthetic peptide: CRQAEGEPSDSSIRRS
anti-NdhB	1:1,000	S. elongatus PCC 7942	directed in this work against synthetic peptide: CLMTGYMKRDPRSNEAALKY
anti-FNR-1	1:1,000	S. elongatus PCC 7942	Directed in this work against synthetic peptide: CRQAEGEPSDSSIRRS
anti-FNR-9a	1:1,000	S. elongatus PCC 7942	Directed in this wok against synthetic peptide: CWSDYQRTLKKAGRWH
anti-Rabbit-HRP	1:2,000	pork	Dako Chemicals, Hamburg, Germany
anti-Rabbit-AP	1:2,000	pork	Dako Chemicals, Hamburg, Germany

The given dilution of the listed antibodies was used, when cell-free French press extracts corresponding to 100 µg protein were subjected to SDS PAGE and immunoblotting.

2.34 Mass spectrometric analysis via MALDI-TOF MS

2.34.1 Coomassie Brilliant Blue stain

For mass spectrometric analysis, PAGEs were stained with Coomassie Brilliant Blue solution (0.08% (w/v) Coomassie Brilliant Blue G250, 1.6% (v/v) H_3PO_4, 8% (w/v) ammonium sulphate, 20% (v/v) methanol) for 24 h. Destaining was performed using H_2O bidest.

2.34.2 Silver stain

For MALDI-TOF compatible silver stain gels were incubated in fixing buffer (50% (v/v) ethanol, 20% (v/v) isopropanol, 0.003% (v/v) formaldehyde). Subsequently, gels were washed twice using washing buffer (50% (v/v) ethanol), followed by an incubation for 1 min in deoxidisation solution (0.02% (w/v) sodium dithionate). Afterwards, gels were washed 3 times with H_2O bidest and incubated in staining solution (0.2% (w/v) $AgNO_3$, 0.02% (v/v) formaldehyde) for 20 min, followed by the development in the developer (12% (w/v) $NaCo_3$, 0.0008% (w/v) sodium dithionite, 0.03% (v/v) formaldehyde) until the staining was accomplished. Afterwards, gels were washed 3 times with H_2O bidest and staining was stopped by washing in stop solution (12% (v/v) acetic acid, 50% (v/v) ethanol). The stained gels were stored at 4°C using 1% (v/v) acetic acid.

2.34.3 Tryptic digest and MALDI-TOF MS

Analysis of protein spots and bands separated via PAGE was performed using *matrix assisted laser desorption/ionisation time of flight mass spectroscopy* (MALDI-TOF MS). This method determines the mass of peptides generated by tryptic digestion after ionisation.

Preparing the tryptic digest 1.5 mL tubes were prewashed 3 times with 500 µL 60% (w/v) acetonitrile and 0.1% (v/v) trifluoroacetic acid. Protein bands were cut out and washed with 250 µL 50% (w/v) acetonitrile for 5 min. After removing the supernatant, 250 µL 50% (w/v) acetonitrile and 50 mM NH_4HCO_3 were applied and samples were incubated for 30 min at room temperature while being slowly shaken. After removing the supernatant 250 µL 50% (w/v) acetonitrile and 10 mM NH_4HCO_3 were added and samples were incubated again for 30 min at room temperature with slow shaking. After removing the supernatant, the pellet was dried in a SPD SpeedVac (Thermo Savant, Thermo Fisher Scientific, Waltham, USA) for 20 min. Subsequently, 10 µL trypsin (10 µg/mL) and 15 µL 10 mM NH_4HCO_3 were given to the dried gel pieces and incubated for 5-10 min at room temperature before adding additional 20 µL 10 mM NH_4HCO_3. Samples were incubated for 24 h at 37°C and stored at 4°C. Protein samples were analysed in a MALDI-TOF MS designed by Bruker (Bruker Daltonic GmbH, Bremen, Germany). The received data was evaluated using the tool *PeptideMass* provided by *Expasy* (http://www.expasy.org).

2.35 Iron determination using ICP-OES

The iron content of heterologous expressed His-tagged IdiC was determined using *inductively coupled plasma optical emission spectroscopy* (ICP-OES). The measurements were carried out in the chair of Prof. Dr. W. Horst for Plant Nutrition at the University of Hanover.

The 6 x His-tagged IdiC protein was expressed in *E. coli* strain BL21 as described in chapter 2.24.1. A cell-free extract was obtained following chapter 2.20.3. The resulting insoluble IdiC inclusion bodies were pelleted by centrifugation at 4°C applying 10,000 x g. The protein pellets were transferred in 1.5 mL test tubes and dried at 60°C using a SPD SeedVac (Thermo Savant, Thermo Fisher Scientific, Waltham, USA). After determination of the corresponding weight, the dried protein pellets were stored at -20°C. The iron content of the defrosted protein was determined after dry ashing at 480°C for 8 h and dissolving the ash in 6 M HCl with 1.5% (w/v) hydroxylammonium chloride (Fecht-Christoffers et al. 2006). Subsequently, this solution was diluted 1:10 with H_2O. Measurements were carried out by optical emission spectroscopy, inductively coupled plasma using a SPECTRO GENESIS spectrometer (Spectro Analytical Instruments GmbH, Kleve, Germany).

2.36 Iron determination using EPR

The iron content of heterologous expressed His-tagged IdiC was additionally determined using *electron paramagnetic resonance spectroscopy* (EPR). The measurements were done in cooperation with the Max Planck Institute for Bioorganic Chemistry located in Mülheim a. d. Ruhr.

The 6 x His-tagged IdiC protein was expressed in *E. coli* strain BL21 described in chapter 2.36. In contrast to the samples used for ICP measurements, IdiC inclusion bodies were solubilised adding 2% (w/v) β-DM. The procedure was performed as described in chapter 2.25.1. The solubilised protein was stored at 4°C. Applying for EPR measurements, the protein solution was transferred to cryotubes and frozen immediately in liquid nitrogen. The EPR emissions spectra were recorded using a Bruker EPR continuous wave X-band spectrometer E500 CW (Bruker Daltonic GmbH, Bremen, Germany). The used EPR parameters included: sample temperature, 14K; microwave frequency, 9.449 GHz; microwave power, 0.2 mW; modulation frequency, 100 kHz; modulation amplitude, 20.0 G; scan field [G], 500-4500; averaged scans, 42; and time constant, 20.5 ms.

2.37 IdiC localisation using immunocytochemistry

Immunocytochemical procedures were done as described previously (Stephan et al. 2000). After growth in regular and iron deficient BG11 medium, *S. elongatus* PCC 7942 cells were harvested by centrifugation (Labofuge Ae, Heraeus Sepatech, 20 min at 4,000 rpm) and fixed for 30 min with 2.5% glutaraldehyde. After dehydration using ethanol, cells were embedded in LR white resin (London Resin Company, Berkshire, UK). The anti-IdiC antiserum used for the detection was raised in rabbit (see chapter 2.26) and diluted 1:100.

As secondary antibody a gold-coupled anti-rabbit IgG was used with a dilution of 1:30. The electron microscopical pictures were recorded by a Hitachi H-500 transmission electron microscope (Hitachi, Tokyo, Japan) applying 75 kV.

(D) Bioinformatic methods

2.38 Database searches and sequence analysis

Cyanobacterial genome sequences were obtained from the CyanoBase database (http://bacteria.kazusa.or.jp/cyanobase/cyano.html) or the JGI microbial genomics database (http://genome.jgi-psf.org/mic_home.html). Database searches and similarity searches were done as described previously (Rueckert et al. 2003) with nucleotide and amino acid sequences using the BlastN- and BlastP-algorithms (http://blast.ncbi.nlm.nih.gov/Blast.cgi) (Altschul et al. 1997). Multiple sequence alignments were performed using the DIALIGN2 software (http://bibiserv.techfak.uni-bielefeld.de/dialign/) (Morgenstern 1999). Illustration of sequence similarities were done with the WebLogo tool (http://weblogo.berkeley.edu/).

Bioinformatic analyses to identify protein parameters and putative domains were performed using the following programs:

    SMART Sequence (http://smart.embl-heidelberg.de/)
    ClustalW2 (EMBL-EBI, http://www.ebi.ac.uk/Tools/clustalw2/index.html)
    Pfam Motif Search (http://pfam.sanger.ac.uk/search)
    TMpred (http://www.ch.embnet.org/software/TMPRED_form.html)
    DAS TM prediction server (http://www.sbc.su.se/~miklos/DAS/)
    InterProScan (EMBL-EBI, http://www.ebi.ac.uk/Tools/InterProScan/)
    ProtParam tool (http://www.expasy.org/tools/protparam.html)
    PeptideMass (http://www.expasy.org/tools/peptide-mass.html)

3 Results

3.1 Bioinformatic analysis of the idiC gene and its gene product

The preliminary characterisation of the *orf6-idiC-idiB* operon, subsequently called *idiB* operon, (EMBL nucleotide database entry Z48754) shows that the coding sequence comprises the following bps: *orf6* bps 5004-4744, *orf5* (representing *idiC*) bps 4747-4187, and *orf4* (representing the *idiB* gene) bps 4090-3464. Thus, the genes *orf6* and *orf5* overlap by 3 bps. For an overview on database entry Z48754 see Figure 3.1. The transcription of the *idiB* operon is induced in the course of increasing iron starvation as well as by oxidative stress (Michel and Pistorius 2004; Yousef et al. 2003). It has been suggested, that at least four different transcripts originate from a primary tricistronic transcript (Yousef et al. 2003).

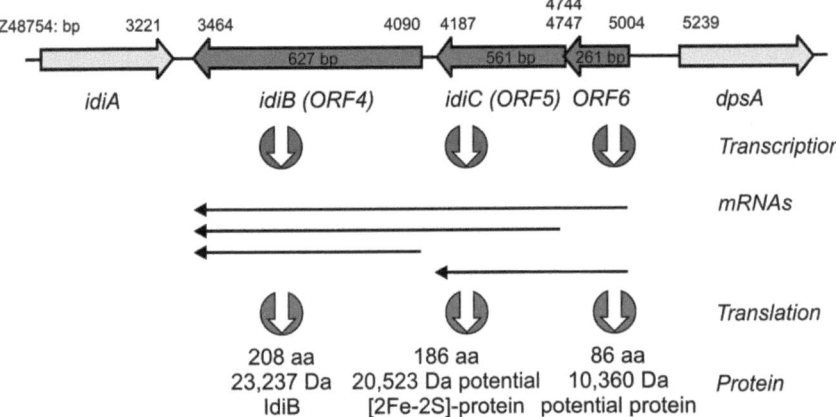

Figure 3.1: Physical map of EMBL database entry Z48754. The 5.8 kb *Hin*dIII fragment of genomic DNA from *Synechococcus elongatus* PCC 6301 carries the *idiA* gene and the genes *idiB* and *dpsA*, whose gene products regulate and influence transcription of *idiA*. The *idiB* operon in *S. elongatus* PCC 7942 consists of the genes *idiB*, *idiC*, and *orf6*. Its primary and secondary transcripts as well as the basics of the putative encoded proteins are shown below (Yousef et al. 2003).

Due to the fact that the *orf5* gene becomes transcribed and translated into a protein, the name *orf5* has been changed into *idiC* for iron deficiency-induced protein C in the database entry Z48754. The deduced amino acid sequence of IdiC contains 186 amino acid residues. IdiC has a molecular mass of 20.523 Da and a calculated pI of 9.17 (ProtParam prediction, http://www.expasy.org/tools/protparam.html).

For the identification of potential transmembrane helices, the transmembrane prediction tools on the ExPASy homepage (http://www.expasy.org/tools/) were used (DAS-TM, ProtTM, PredictProtein, TopPred, SODUI). All tools classified IdiC to be a soluble protein without any lipophilic domains. The DAS prediction tool (ExPASy) detects important regions of hydrophobic amino acids as putative transmembrane regions (loose cutoff), but even this software did not reveal the presence of any putative transmembrane region. The corresponding DAS prediction spectrum is shown in Figure 3.2.

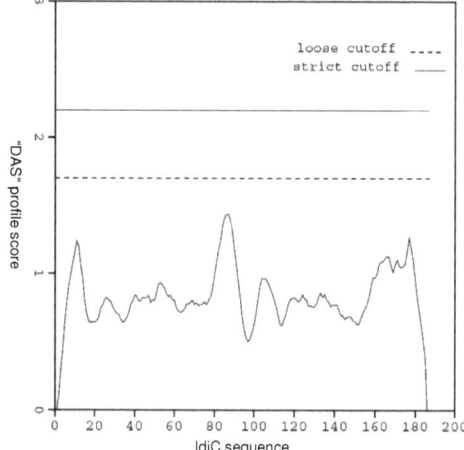

Figure 3.2: Evaluation of the putative IdiC amino acid sequence, obtained by the DAS transmembrane prediction tool at ExPASy (http://www.sbc.su.se/~miklos/DAS). Since the peaks indicating transmembrane helices do not even rise above the loose cutoff threshold, IdiC most likely is a soluble protein without any transmembrane regions.

Similarity and domain searches (BlastP search at NCBI, InterPro Scan at EMBL) indicated that IdiC belongs to the family of thioredoxin (TRX)-like [2Fe-2S] ferredoxins. This protein family is composed of [2Fe-2S] ferredoxins with a TRX-fold (TRX-like ferredoxins) and proteins containing domains similar to TRX-like ferredoxins such as e.g. the NADH:ubiquinone oxidoreductase (NDH-1) subunits NuoE and NuoF of *E. coli*. An alignment of the predicted [Fe-S]-binding site of IdiC with the cysteine-containing [Fe-S]-binding motif of ferredoxin PetF2 from *Synechococcus* sp. PCC 7002 (Pfam Motif Search), the subunits NuoE and NuoF of bacterial NADH:ubiquinone oxidoreductases (Friedrich 1998) as well as with the subunit HoxF of the hydrogenase-diaphorase complex (Friedrich and Scheide 2000) is presented in Figure 3.3. The similarity is predominantly related to the binding site of the [2Fe-2S] cofactor. The four cysteine residues, which are supposed to be directly involved in binding of the cofactor, are marked in yellow boxes.

```
IdiC   S. PCC 7942   AVLEVCTRGTCRRRGALELCDRLQA----EAKTC----AVEVRTRGCLGRCKQGINVRRS
PetF2  S. PCC 7002   ----ICCHHTCPKQGSTAILAAFQA----QAPA-----DVEVRQAGCFGECGNGPLVR--
PetF   H. NRC-1      ----VCTNQTCAAEGAPAVLERLRQ----EARDADE-DSLRVTRTSCLGQCGDGPNVAVY
NuoE   E. coli       HHIRVCTGTACHLKGSEALLKALEKKLGIKPGETTADGKFTLEPVECLGACGQAPVVMIN
NuoF   B. VPI-5482   ----ICGGTGCKASSSQGITENLQK----AIERNGITDKVDVITVGCFGFCEKGPIVKII
HoxF   S. PCC 6301   IRLRCCTATGCRANGAEAVFKAVQQ----TIADQNLGDRCEAVSVGCLGLCGAGPLVQCD
                        *     *         .:     :    ..                  *:*  *   . *
```

Figure 3.3: Comparison of a partial IdiC protein sequence from *S. elongatus* PCC 7942 and selected proteins with a [2Fe-2S]-binding site using the ClustalW software at EMBL (http://www.ebi.ac.uk/clustalw/): NuoE from *E. coli* strain K12 (NP_416788); NuoF from *Bacteroides thetaiotaomicron* VPI-5482 (AAO75232); PetF2 from *Synechococcus* sp. PCC 7002 (AAN03538); PetF from *Halobacterium* sp. NRC-1 (NP_444222); HoxF from *Synechococcus elongatus* PCC 6301 (CAA73873). The four cysteine residues involved in binding of the [2Fe-2S] cluster are boxed with a solid line. * indicate identical amino acid residues, : indicate conservative amino acid substitutions (A/V/F/P/M/I/L/W, D/E, R/H/K, S/T/Y/H/C/N/G/Q, and X indicate semi-conserved amino acid residues. Gaps were inserted into the sequences to maintain optimal alignment.

The highest similarity of the full-length IdiC protein exists to NuoE of *E. coli* strain K12. The IdiC protein contains 46% similar amino acid residues, including 16% identical residues. An alignment of these two proteins is given in Figure 3.4. NuoE belongs to the peripheral subunits and represents the substrate-binding part of the bacterial NDH-1 complex of *E. coli*. It contains a [2Fe-2S] centre and has a molecular mass of 18.5 kDa (Friedrich and Weiss 1997).

```
IdiC    MKLPFYLEGYFLGLQDPDTPDNIRFIHVQSD-GGNRYTLKLAKPLRHLPWQQLTVGQPLR  59
NuoE    ------------MHENQQPQTEAFELSAAEREAIEHEMHHYEDPRAASIEALKIVQKQR  47
        : : :  *:.   *     ::  . .: ::   :  *  . : *.:  * *

IdiC    -IEGQQSFQGLDLPPKLKAERVLFDPAGLPAFIASEEPPKPQPAVLEVCTRGTCRRRGAL 118
NuoE    GWVPDGAIHAIADVLGIPAS----DVEGVATFYSQIFRQPVGRHVIRYCDSVVCHINGYQ 103
        :  :::.:         : *.    *  *:.:*  :.        *:. *  .*:  .*
IdiC    ELCDRLQAEAKTCAVEVRTRG-------CLGRCKQGINVRRSSDNQILSQLSPQAAAEL  170
NuoE    GIQAALEKKLNIKPGQTTFDGRFTLLPTCCLGNCDKGPNMMIDEDT--HAHLTPEAIPEL 161
        :  *: :  :  . :.    *        ***.*.:* *:   ..* .  ::*:*:* .**

IdiC    LSPWRTPAVVSGTAVG 186
NuoE    LERYK---------- 166
        *. ::
```

Figure 3.4: Alignment of IdiC from *S. elongatus* PCC 7942 and NuoE from *E. coli* K12 (NP_416788). Amino acid similarity corresponds to 46% including 16% identity. IdiC has a total length of 186 amino acid residues which corresponds to a molecular mass of 20.5 kDa. NuoE has a total length of 166 amino acid residues which corresponds to a molecular mass of 18.5 kDa. * indicate identical amino acid residues, : indicate conservative substitutions (A/V/F/P/M/I/L/W, D/E, R/H/K, S/T/Y/H/C/N/G/Q, and X indicate semi-conserved amino acid residues. Gaps were inserted into the sequences to maintain an optimal alignment.

Performing a genomic BlastP search with the deduced amino acid sequence of IdiC at the NCBI homepage (http://blast.ncbi.nlm.nih.gov/Blast.cgi), orthologues of IdiC were found in the genomes of several other cyanobacterial species. Moreover, these results revealed a strong similarity to a great number of predicted cyanobacterial NDH-1 subunits. Since these proteins in general contain more than 500 aa residues and the similarity of IdiC is strictly limited to the [Fe-S]-binding site, the overall *e*-value of the corresponding blast searches is rather low. In the group of fresh water cyanobacteria, the majority of mesophilic strains and all thermophilic strains contain a gene encoding an IdiC-similar protein. An exception is *Synechocystis* sp. PCC 6803, which does not contain such a protein. In contrast to fresh water cyanobacteria, the so far sequenced genomes of the marine *Prochlorococcus* strains as well as the marine *Synechococcus* strains do not possess a gene encoding an IdiC-similar protein.

The characteristics of the cyanobacterial IdiC-similar proteins as well as their identity and similarity values to IdiC from *S. elongatus* PCC 7942 are presented in Table 3.1. All query proteins belong to the thioredoxin (TRX)-like [2Fe-2S] ferredoxin family and are composed of [2Fe-2S] ferredoxins with a TRX-fold including formate dehydrogenases, NAD-reducing hydrogenases, and the subunit E of the bacterial NADH:ubiquinone oxidoreductase NuoE. The finding that IdiC has a high similarity to NuoE is particularly interesting, since so far only 11 genes with similarity to the 14 genes encoding subunits of the NDH-1 complex in *E. coli* have been identified in cyanobacteria (Berger et al. 1993; Friedrich and Scheide 2000; Friedrich and Weiss 1997).

Table 3.1: Results of a NCBI genomic BlastP search for cyanobacterial proteins with similarity to the IdiC protein from S. elongatus PCC 7942 in so far sequenced 33 cyanobacterial genomes at NCBI (March 2007). Identity and similarity values correspond to full length sequences of IdiC and query proteins. All these proteins belong to the family of [2Fe-2S] ferredoxins with a TRX fold including formate dehydrogenases, NAD-reducing hydrogenases, and the substrate binding subunit of the bacterial NADH:ubiquinone oxidoreductase NuoE (Pietsch et al. 2007). "SU" subunit.

Protein	Organism	Function	e-value	Identity (%)	Similarity (%)	AA	MM	pI	Database accession No.
2173	S. elongatus PCC 7942	Putative [Fe-S] protein	2e-106	100	100	186	20.5	9.17	YP_401190
Syc_1922	S. elongatus PCC 6301	Putative [Fe-S] protein	2e-106	100	100	186	20.5	9.17	YP_172632
2462	S. elongatus JA-3-3AB	Putative [Fe-S] protein	5e-12	31	46	184	20.2	9.66	YP_475847
5165	Crocosphera watsonii WH 8501	Putative [Fe-S] protein	4e-10	28	46	185	21.1	9.72	ZP_00514983
08091	Lyngbia sp. PCC 8106	Putative [Fe-S] protein	8e-08	27	46	189	20.9	9.51	ZP_01624781
Alr3959	Anabaena variabilis ATCC29413	Putative [Fe-S] protein	1e-07	28	42	188	20.6	9.12	YP_324459
Alr1410	Nostoc sp. PCC 7120	Putative [Fe-S] protein	1e-07	29	42	188	20.7	9.17	NP_485453
CYB_1432	S. elongatus JA-2-3Ba (2-13)	Putative [Fe-S] protein	2e-07	27	44	216	23.2	9.20	YP_477662
4529	Anabaena variabilis ATCC29413	Putative [Fe-S] protein	3e-07	28	43	219	24.2	9.21	YP_325022
All0596	Nostoc sp. PCC 7120	Putative [Fe-S] protein	1e-06	27	44	219	24.3	9.28	NP_484640
Tlr1919	Thermosynechococcus elongatus BP-1	Putative [Fe-S] protein	8e-06	29	45	172	19.3	9.69	NP_682709
4280	Trichodesmium erythraeum IMS 101	Putative [Fe-S] protein	1e-05	23	40	222	24.8	9.69	YP_723750
Npun02003199	Nostoc punctiforme PCC 73102	24 kDa SU bidirectional hydrogenase	3e-05	28	44	218	24.1	9.46	ZP_00109566
Gll2565	Gloeobacter violaceus CC 7421	Putative [Fe-S] protein	5e-05	31	43	199	22.1	9.04	NP_925511
9761	Nodularia spumigena CCY 9414	Putative [Fe-S] protein	8e-05	28	45	218	24.1	9.07	ZP_01632334
NPpun02006394	Nostoc punctiforme PCC 73102	Putative [Fe-S] protein	1e-03	28	38	180	19.8	9.35	ZP_00106962

No genes with similarity to the three peripheral subunits of the substrate-binding part of the bacterial NDH-1 complex, named NuoE, F, and G, have so far been identified. Thus, the direct substrate for the cyanobacterial NDH-1 complex as well as its substrate-binding subunits have remained unidentified (Friedrich and Weiss 1997; Vermaas 2001). Based on the results presented in this work, the NuoE-similar IdiC may be a likely candidate for one of the so far unidentified subunits for the substrate-binding part of the cyanobacterial NDH-1 complex.

3.2 Expression of IdiC in *S. elongatus* PCC 7942

For investigation of the IdiC expression pattern in *S. elongatus* PCC 7942 WT, cells were grown for different times in iron-sufficient or iron-deficient BG11 medium and were harvested for isolation of total RNA and preparation of cell-free French press extracts. The *idiC* transcription and translation was detected by Northern and immunoblot experiments as well as by immunocytochemistry.

3.2.1 Expression of IdiC under iron-deficient growth conditions

For characterisation of the expression of IdiC in *S. elongatus* PCC 7942 WT, cells were inoculated with an optical density at 750 nm of 0.4 and cultivated for 48, 72, and 96 h under iron-sufficient and iron-deficient growth conditions. Subsequently, cells were harvested for isolation of total RNA and preparation of cell-free extracts.

Northern blot analyses showed that the steady-state transcript level of *idiC* mRNA was highly up-regulated under iron limitation (see Figure 3.5 A). This result has been anticipated, since the entire *idiB* operon is known to be iron-regulated (Yousef et al. 2003). Additionally, the *idiC* transcript pool became up-regulated after prolonged growth under iron-sufficient growth conditions. After 96 h the content of the *idiC* mRNA was almost similar to that of cells grown under iron-deficient conditions. 10 µg total RNA isolated from *S. elongatus* PCC 7942 WT cells were used for the detection of the *idiC* transcript by hybridisation with a Dig-dUTP-labelled *idiC*-specific probe using CDP-Star™ detection.

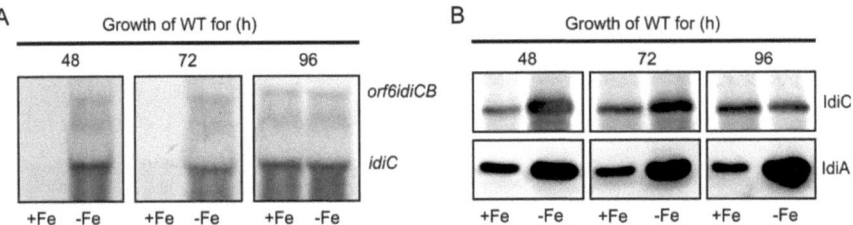

Figure 3.5: Investigation of *idiC* transcription and translation in *S. elongatus* PCC 7942 after growth with iron-sufficient and iron-deficient BG11 medium for 48, 72, and 96 h. **(A)** Northern blot of the *idiC* expression in *S. elongatus* PCC 7942 WT cells. The *idiC* transcription was up-regulated under iron starvation and after prolonged growth under iron-sufficient conditions. The *idiC* transcript was detected using a gene-specific Dig-dUTP-labelled probe. **(B)** Immunoblot detection of IdiC and IdiA expression in *S. elongatus* PCC 7942. After growth for 96 h, the IdiC content was almost similar in cells grown under iron-sufficient and iron-deficient conditions, while the IdiA content was further increased in cells grown under iron-deficient conditions as compared to cells grown under iron-sufficient conditions.

Immunoblot analysis with the anti-IdiC antiserum provided evidence that in addition to the *idiC* transcript the IdiC content was also strongly increased in *S. elongatus* PCC 7942 WT cells under iron-starved conditions (see Figure 3.5 B). For immunoblot analysis cells were harvested and broken in a French Press. The amount of IdiC and IdiA protein was detected by immunoblotting with the anti-IdiC and the anti-IdiA antiserum. 100 µg of protein were applied to SDS PAGE for detection with the anti-IdiC antiserum (dilution 1:300) and 50 µg of protein were applied using the anti-IdiA antiserum (dilution 1:5,000). Development of blots was performed with the ECL™ detection kit (GE Healthcare).

In agreement with the Northern blot results, the immunoblot investigations provided evidence that the IdiC content increased after growth for 96 h with regular BG11 medium. It became obvious, that in the late growth phase practically no difference in the IdiC content of *S. elongatus* PCC 7942 WT cells grown with regular BG11 medium and cells grown in BG11 medium from which iron was omitted was present. However, the IdiA content was further increased in cells grown under iron-deficient conditions as compared to cells grown under iron-sufficient conditions.

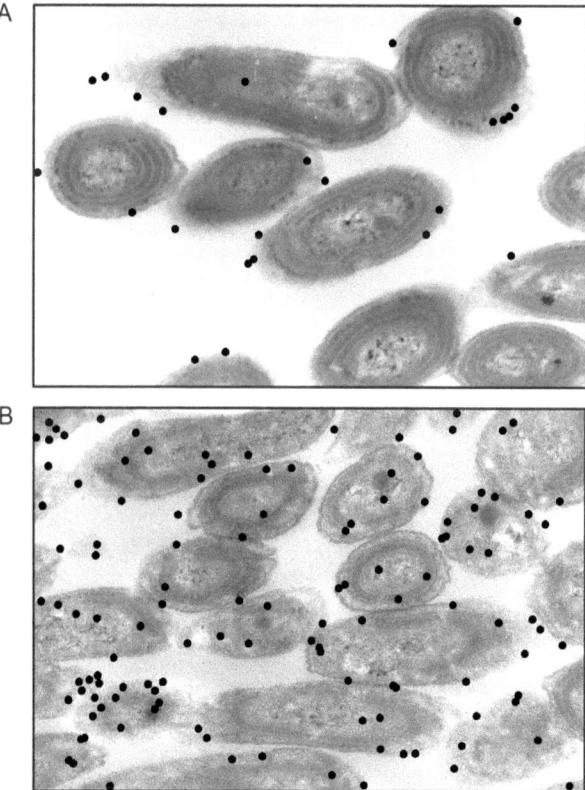

Figure 3.6: Electron micrographs of *S. elongatus* PCC 7942 WT cells grown for 24 h either under iron-sufficient (42,000-times magnified) **(A)**, or under iron-deficient conditions (32,000-times magnified) **(B)**. The immunostaining was performed with the anti-IdiC antiserum (dilution 1:300) and shows the increased expression of IdiC in iron-starved cells. The gold labels were recoloured in yellow to improve visibility.

Immunocytochemical investigations of intact S. elongatus PCC 7942 WT cells confirmed the results obtained by immunoblotting. Cells grown for 24 h in regular BG11 medium were only slightly labelled, indicating a low intracellular IdiC content (see Figure 3.6 A), whereas S. elongatus PCC 7942 WT cells grown for the same time under iron-deficient conditions showed a significantly increased labelling of IdiC proteins (see Figure 3.6 B). Thus, it has been demonstrated by two different methods that the IdiC expression in S. elongatus PCC 7942 WT depends on the cellular iron status and is strongly increased under iron-starved growth conditions.

3.2.2 Expression of IdiC in the late growth phase

For investigation of the IdiC expression during prolonged growth under iron-sufficient conditions, WT cells were grown in regular BG11 medium and harvested after 24, 48, 72, 96, 120, and 144 h. 50 µg protein of the corresponding cell-free extracts were applied to SDS PAGE and subsequently to immunoblotting with anti-IdiC and anti-IdiA-specific antisera.

The results provided evidence that cells, which were grown with regular BG11 medium for 144 h, contained a significantly higher level of IdiC protein than cells harvested during the early growth phase. Similarly, the IdiA content increased in cells grown for 144 h under iron-sufficient conditions (see Figure 3.7 A), indicating a reduction of free iron concentration in the growth medium or an increasing oxidative stress. To decide whether IdiC expression is caused by a reduction of iron concentration in the medium in the late growth phase, additional 30 µM ferric citrate (FeCitrate) were added to the cultures after 72 h of growth. Figure 3.7 B shows that the addition of iron led to a minor reduction of IdiA in cells harvested after 144 h. Intriguingly, the treatment did not result in a reduction of IdiC concentration, but led to a further increased IdiC content as compared to cells grown for 144 h without additional ferric iron.

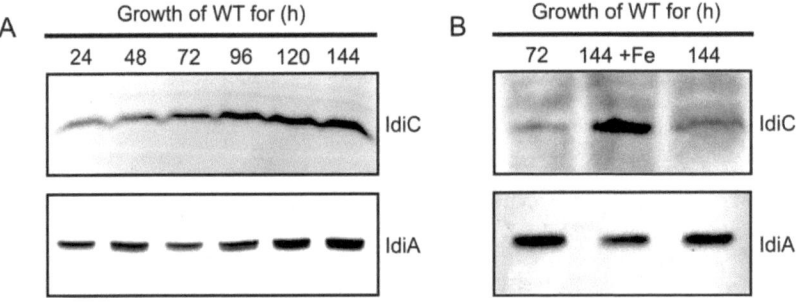

Figure 3.7: Time course of IdiC and IdiA expression under iron-sufficient growth conditions in S. elongatus PCC 7942 WT cells. Immunoblots were performed with the anti-IdiC (dilution 1:300) and the anti-IdiA antiserum (dilution 1:2,000). **(A)** Cells were grown with regular BG11 for times indicated (24-144 h). Prolonged growth under iron-sufficient growth conditions leads to an increased IdiC and IdiA expression. **(B)** WT was grown for 144 h in regular BG11 medium to which additional 30 µM Fecitrate were added after 72 h. After 144 h, the iron-dependent IdiA expression was slightly decreased, while the IdiC content of cells grown in the replenished medium was higher than in cells grown with BG11 medium without additional iron.

The results obtained by immunoblot analysis were approved by immunocytochemical investigations illustrated in Figure 3.8. Cells were harvested after growth for 24 and 144 h and subsequently fixed in LR white resin (London Resin Company). The IdiC protein was detected with the anti IdiC-antiserum. *S. elongatus* WT cells grown for 24 h under iron-sufficient conditions (see Figure 3.8 A) revealed only a minor number of gold labels compared to those in Figure 3.6 B. In contrast, cells grown for 144 h in regular BG11 medium showed a high number of gold-labelled second antibody detecting the IdiC protein expression via the anti-IdiC antiserum (see Figure 3.8 B).

Therefore, it is highly likely that IdiC is indeed a protein, which becomes expressed in elevated amounts in the late growth phase of *S. elongatus* PCC 7942.

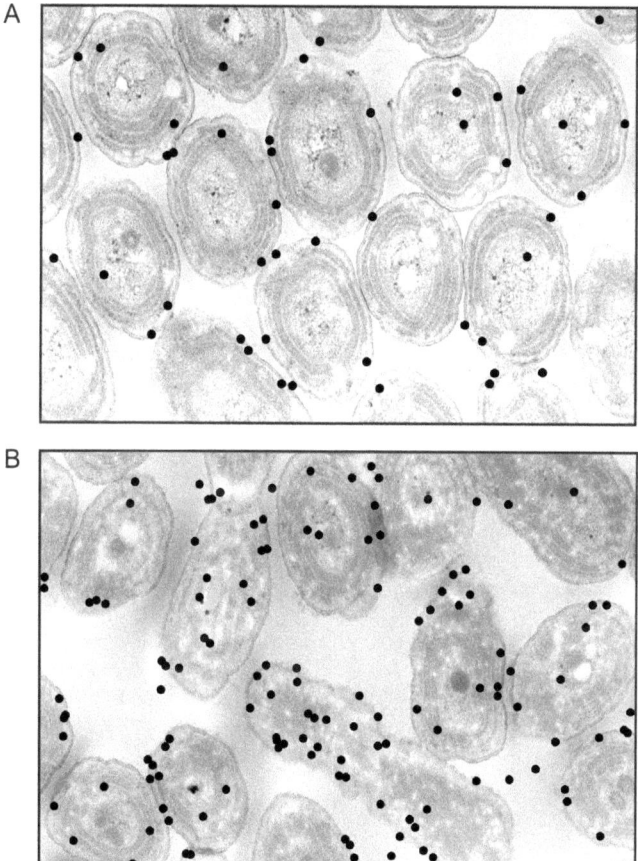

Figure 3.8: Expression of IdiC in the late growth phase. The IdiC protein was detected with the anti-IdiC antiserum diluted 1:300. **(A)** WT cells (32,000-times magnified) grown for 24 h in regular BG11 medium showed only a weak labelling of IdiC. **(B)** The expression of IdiC increased significantly in cells harvested after a growth of 144 h under iron-sufficient conditions (37,000-times magnified). The immunogold labels were restained in yellow to improve the visibility.

3.3 Investigation of the localisation of IdiC in *S. elongatus* PCC 7942

For further elucidation of the function of the IdiC protein, extended localisation studies in *S. elongatus* PCC 7942 WT were performed. Bioinformatic analyses of the deduced amino acid sequence showed that IdiC contained no lipophilic transmembrane regions (TM prediction tool) and thus, was classified to be a soluble protein (see chapter 3.1). The first investigations were focused on soluble cell fractions. The results provided evidence that IdiC is loosely attached to the thylakoid membrane. Investigations were carried out by immunoblot analyses of subcellular fractions, immunocytochemical investigations, BN PAGE and isolation of PS I complexes.

3.3.1 Localisation of IdiC in subcellular fractions of *S. elongatus* PCC 7942

For investigation of subcellular fractions *S. elongatus* PCC 7942 WT, cells were cultivated for 48 h in BG11 from which iron was omitted. Subsequently, cells were harvested for preparation of soluble periplasmic proteins and the corresponding spheroplasts (see chapter 2.5.2). The periplasmic fraction was concentrated by dialysis against Carbowax™ (Sigma). Spheroplasts were broken in a pre-chilled French press cell to release the soluble and insoluble spheroplast proteins. Periplasmic and spheroplast protein fractions corresponding to 100 µg protein were subjected to SDS PAGE and subsequently to immunoblotting. Immunoblots were performed with the anti-IdiC antiserum at a dilution of 1:300. The IdiC protein was exclusively detected in the spheroplastic fraction indicating that IdiC is located intracellularly (see Figure 3.9 A).

To examine the intracellular localisation of IdiC in greater detail, cells were harvested after 48 h of iron-deficient growth for preparation of cell-free extracts. Cells were resuspended in buffer containing 50 mM HEPES pH 6.5, 0.05 M $CaCl_2$ and 0.4 M sucrose to prevent decay of membrane-protein complexes during isolation of the membrane fraction and the corresponding soluble periplasmic proteins by centrifugation at 12,000 x g. Immunoblots with the anti-IdiC antiserum were carried out as described above. Figure 3.9 B shows, that the IdiC protein was detected in significant amounts in the membrane fraction. In contrast, only a minor amount was detected in the soluble fraction. This result suggested that IdiC is loosely attached to the cytoplasmic and/or the thylakoid membranes.

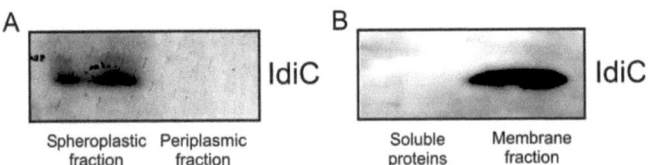

Figure 3.9: Immunoblots of subcellular fractions of *S. elongatus* PCC 7942 WT cells grown for 96 h in iron-deficient BG11 medium. 100 µg protein were subjected to SDS PAGE. **(A)** Spheroplastic and the corresponding periplasmic fraction prepared as described in the literature (Block and Grossmann 1988). **(B)** Intracellular soluble and membrane fraction isolated using sucrose-containing buffer. The IdiC protein is located intracellularly and exclusively found in the membrane fraction.

Membrane fractions isolated as described above and shown in Figure 3.21 B predominantly contained mostly thylakoid membranes and only minor amounts of cytoplasmic membrane. To investigate whether IdiC is attached to the cytoplasmic or to the thylakoid membrane, the different types of membranes were separated by density gradient centrifugation as described by Omata and Murata (1984). *S. elongatus* PCC 7942 WT cells were cultivated for 144 h in regular BG11 to which additional FeCitrate was added after 72 h. Cells were harvested, resuspended in sucrose-containing Tes-NaOH pH 6.5, treated with lysozyme, and broken by passage through a pre-chilled French pressure cell. The cell-free extract was supplied to a discontinuous sucrose gradient of 10 to 90%. The different membrane systems were subsequently separated via centrifugation at 130,000 x g and distributed in the sucrose gradient as illustrated in Figure 3.10 A. The separated membranes were withdrawn from the gradient and, if necessary, concentrated via centrifugation using Centricon™ concentrators (Millipore). 100 µg proteins of each fraction were subjected to SDS PAGE and immunoblotting. IdiC detection with the anti-IdiC antiserum was done as described above.

Figure 3.10 B shows that the IdiC protein was present in the soluble fraction containing high concentrations of phycobilisome proteins. Only a minor IdiC content was detected in the thylakoid membrane fraction. Due to the results obtained from thylakoid membrane isolation illustrated in Figure 3.21 B, these results came rather unexpected and thus, indicated that the IdiC protein was separated from the membrane during sucrose gradient centrifugation at 130,000 x g.

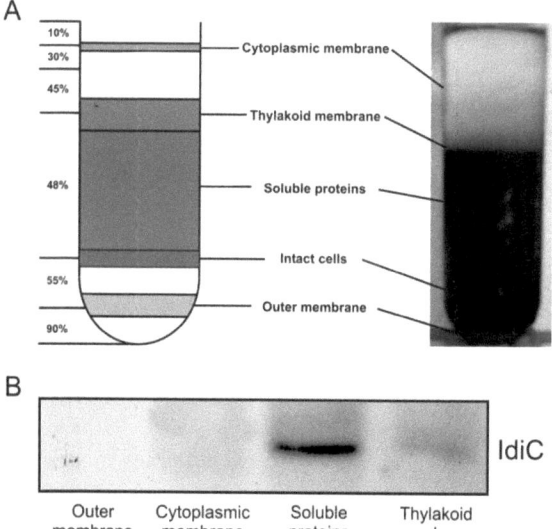

Figure 3.10: Isolation of membrane fractions of *S. elongatus* PCC 7942 cells by sucrose density centrifugation (Omata and Murata 1984). **(A)** Sucrose gradient from 10% to 90% showing the isolated membranes in comparison to a schematic presentation. **(B)** Immunoblot analysis of the separated and withdrawn membrane systems as well as the corresponding soluble proteins. 100 µg protein were subjected to SDS PAGE and immunoblotting. Dilution of the anti-IdiC antiserum was 1:300. The IdiC protein was detected predominantly in the soluble fraction, whereas the thylakoid membranes contained IdiC only to a minor extent.

Considering the results obtained from both, the thylakoid isolation and the membrane separation by sucrose density gradient centrifugation, IdiC is loosely attached to the thylakoid membranes in S. elongatus PCC 7942.

3.3.2 Immunocytochemical studies on the localisation of IdiC

Previous investigations of subcellular fractions obtained from S. elongatus PCC 7942 WT showed conflicting results detecting IdiC in the thylakoid membranes when isolated by centrifugation and detecting IdiC in the soluble fraction when membranes were separated by sucrose density centrifugation. To clarify the localisation of IdiC, immunocytochemical studies were performed. Cells were cultivated for 144 h in regular BG11 medium to which an additional 30 µM FeCitrate was added after 72 h. After cell harvest, cells were immobilised in LR white resin, and the IdiC protein was detected with the anti-IdiC antiserum. The electron micrograph in Figure 3.11 shows a significant gold-labelling of the thylakoid membranes and thus, confirmed the results documented in chapter 3.3.1. Therefore, IdiC is suggested to be a soluble protein, which is loosely attached to the thylakoid membrane.

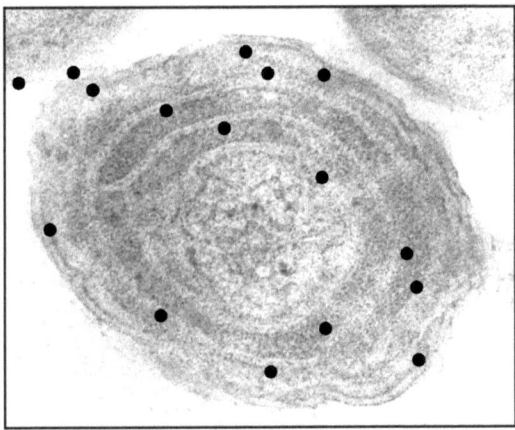

Figure 3.11: Electron micrograph of a S. elongatus PCC 7942 WT cell grown for 144 h in regular BG11 with addition of an extra 30 µM FeCitrate after 72 h (60,700-times magnified). Detection of IdiC with the anti-IdiC antiserum was performed at a dilution of 1:300. Gold labels were redrawn in yellow to improve visibility.

3.3.3 Investigation of thylakoid membranes by BN PAGE

The so far presented results provided evidence that IdiC is attached to the thylakoid membrane or one of its integral membrane protein complexes. For further investigation of the localisation of IdiC in S. elongatus PCC 7942, the multi protein complexes located in the thylakoid membranes were solubilised and separated under native conditions in the first dimension by BN PAGE. The complexes were resolved into individual polypeptides in a second Tris-glycine SDS PAGE dimension. WT cells were grown for 168 h under iron-sufficient conditions. Following the harvest, cells were resuspended in 50 mM HEPES pH 7.0 containing 10% glycerol, followed by preparation of cell-free extracts using the ribolyser (Hybaid). Thylakoid membranes were separated by centrifugation at 15,000 x g and the membrane proteins were solubilised with 1.5% β-dodecyl maltoside (β-DM). For the first gel dimension, solubilised thylakoid fractions corresponding to 1 mg protein were mixed with

Serva Blue G™ (Serva) and subjected to continuous gradient acrylamide gels and subsequently separated by SDS PAGE. Afterwards, lanes of the first dimension gel were cut out and resolved in the second dimension via SDS PAGE. Proteins separated by the first and the second dimension were blotted and subsequently the PsaAB, D1 and IdiC protein were detected with the corresponding antisera and identified by MALDI-TOF-MS analysis.

Figure 3.12 B shows the results obtained by immunoblot analyses of thylakoid membrane protein complexes separated by the first BN PAGE dimension. Solubilised thylakoid membranes corresponding to 250 µg protein were subjected to BN PAGE and immunoblotting. In addition to the main subunits of PS II and PS I, PsaAB and D1, IdiC was also detected in the thylakoid membranes. The corresponding IdiC protein band did not show a size of 21 kDa, but revealed a much higher molecular mass indicating that IdiC might form homo- or heterooligomeric protein complexes with other protein complexes localised in the thylakoid membrane. Since the IdiC band seemed to be localised in the same region like the PsaAB band, this could refer to an association with PS I complexes. The faint lower band detected by the anti-IdiC antiserum was consistent with the deduced molecular mass of IdiC.

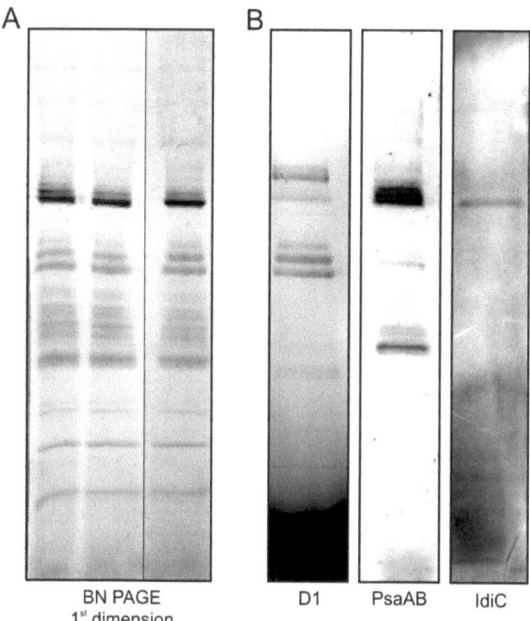

Figure 3.12: BN PAGE of solubilised *S. elongatus* PCC 7942 thylakoid membranes. Cells were grown for 144 h in regular BG11 with addition of an extra 30µM FeCitrate after 72 h. **(A)** Unstained BN PAGE after electrophoresis at 4°C. **(B)** Corresponding immunoblots with the anti-D1 antiserum (dilution 1:2,000), the anti-PsaAB antiserum (dilution 1:1,000), and the anti-IdiC antiserum (1:500). 250 µg protein were subjected to BN PAGE and immunoblotting. IdiC was detected in the thylakoid membrane. The determined molecular mass of the IdiC protein band suggested that IdiC may form complexes with other proteins located in the thylakoid membranes.

Due to the results obtained from the first BN PAGE dimension, it was investigated whether the IdiC protein is connected to a subunit of one of the protein complexes embedded in the thylakoid membrane. The second gel dimension was carried out as Tris-Glycin SDS PAGE resolving the protein complexes into the corresponding subunits. The polypeptides were blotted onto nitrocellulose membranes and detected by immunoblotting using the anti-D1, the anti-PsaAB, and the anti-IdiC antisera as described above. Blots were developed using the ECL™ detection kit (GE Healthcare) and subsequently stained with Ponceau Red (not shown). Comparing the protein spots visible on the developed X-ray films (not shown) to the protein pattern of the stained blots, the appropriate protein spots were detected on the blot membrane. The corresponding spots were cut out from a newly performed silver stained second dimension, trypsin-digested, and subsequently analysed by MALDI-TOF MS. Analysed spots were framed in Figure 3.13. The IdiC spot was confirmed by MALDI-TOF MS and found in the same lane with the PS I reaction centre subunit II, which corresponds to PsaD, which was also identified by MALDI-TOF MS. Thus, parts of the IdiC pool were found to be connected to subunits of PS I under the used growth conditions.

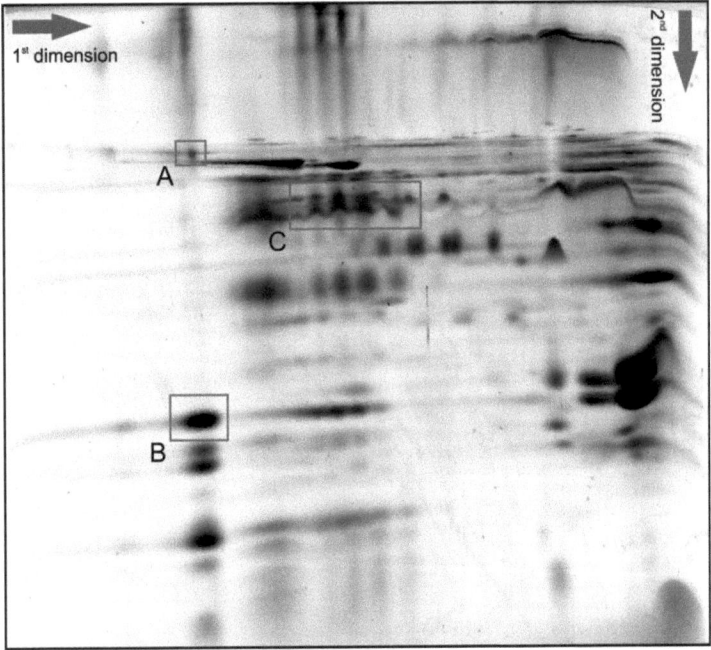

Figure 3.13: Two-dimensional PAGE of solubilised *S. elongatus* PCC 7942 thylakoid membranes. β-DM-solubilised thylakoid membranes corresponding to 1 mg protein were applied to BN PAGE in the first dimension and afterwards applied to regular SDS PAGE in the second dimension. The gel was silver stained and selected protein spots were identified by MALDI-TOF MS. **(A)** Identification of IdiC by immunoblotting and MALDI-TOF MS. **(B)** Identification of PsaD by MALDI-TOF MS. **(C)** PS II subunits identified by immunoblotting. The IdiC protein was identified to run in the same lane as PsaD of PS I. Dilution of antisera corresponds to that given in the legend of Figure 3.24.

3.3.4 Purification and investigation of trimeric PS I complexes

BN PAGE analyses of solubilised thylakoid membranes provided evidence that IdiC might be associated with PS I. For further investigations, trimeric PS I particles were purified from solubilised S. elongatus PCC 7942 thylakoid membranes according to El-Mohsnawy (2007). Cells were grown for 144 h with regular BG11 medium to which an extra 30 µM FeCitrate was added after 96 h. Cells were harvested and resuspended in 20 mM Mes pH 6.5 containing 0.5 M D-mannit. Subsequently, cell-free extracts were prepared and thylakoid membranes were prepared by centrifugation at 100,000 x g. Thylakoid membranes were washed and solubilised using β-DM. A Chl a content of 50% compared to that of the corresponding thylakoid membranes should be solubilised to accumulate primarily soluble PS I complexes. PS I particles were purified from the solubilised thylakoids via HPLC using two chromatography steps. Solubilised thylakoids were applied to HIC as first chromatographic step followed by ion exchange chromatography (IEC) as second step. Fractions resulting from both chromatography steps were concentrated, desalted, and subsequently examined by immunoblot analyses. The work was carried out in cooperation with Dipl.-Biol.'in Marta J. Kopczak in the lab of Prof. Dr. Mathias Rögner at the Ruhr-University Bochum.

3.3.4.1 Purification of PS I by HIC

Thylakoid membranes were extracted from S. elongatus PCC 7942 WT cells harvested after 144 h of growth with regular BG11 medium, which was supplemented with an extra 30 µM FeCitrate after 96 h. Membrane proteins were solubilised by addition of 0.5% β-DM. The solubilised proteins were separated by centrifugation and the main subunits of PS I and PS II as well as IdiC were detected by immunoblotting with the anti-PsaAB, the anti-D1, and the anti-IdiC antiserum. Figure 3.14 shows that both photosystems are mainly solubilised and detectable in the soluble as well as in minor amounts in the insoluble membrane fraction. In contrast, IdiC was entirely solubilised and therefore, barely detectable in the insoluble fraction.

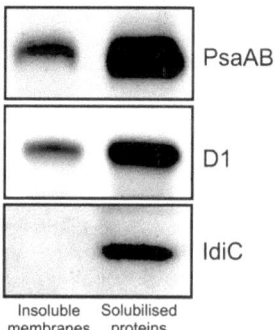

Figure 3.14: Immunoblot analysis of the thylakoid solubilisation using 0.5% β-DM. 50 µg protein were subjected to SDS PAGE and blotting. Detection was performed with the anti-PsaAB (dilution 1:1,000), the anti-D1 (dilution 1:2,000) and the anti-IdiC antiserum (dilution 1:300). PS I and PS II complexes were mainly solubilised, whereas IdiC was detected entirely in the soluble fraction.

Solubilised thylakoid membranes were loaded onto an equilibrated POROS 50 OH column (Applied Biosystems). After a washing step with high salt buffer, PS I particles were eluted with a $(NH_4)_2SO_4$ gradient from 1.5 to 0 M. The detailed HPLC program for the first chromatography is given in the supplementary files. This method was developed for preparation of trimeric PS I particles from thylakoids of Thermosynechococcus elongatus BP-1, subsequently referred to as T. elongatus. Figure 3.15 shows the hydrophobic interaction chromatography (HIC) elution profile - obtained by purification of S. elongatus PCC 7942 PS I particles; and the corresponding conductivity, compared to an HIC profile obtained during PS I preparation from T. elongatus thylakoids (El-Mohsnawy 2007). The first peak of the T. elongatus HIC profile resembled PS II particles, which were already eluted from the column at high $(NH_4)_2SO_4$ concentrations. The next peaks showed that PS I monomers represented by the second peak were eluted previous to PS I dimers and particularly PS I trimers. Compared to the T. elongatus elution profile, the S. elongatus PCC 7942 profile showed that the third peak, representing PS I trimers and dimers, is much lower while the content of PS I monomers is highly elevated compared to T. elongatus. Thus, the results indicated that the protein complexes located in the thylakoid membrane of S. elongatus were much more unstable under the used conditions than those in T. elongatus. In contrast to PS I trimer fractions obtained from T. elongatus, the third HIC fraction corresponded to very low concentrated particles. In addition, it was coloured in deep blue indicating a high content of phycobilisomes. Thus, the fraction, which should have corresponded to T. elongatus PS I monomers, was concentrated, desalted, and subsequently used for the second chromatography step via IEC.

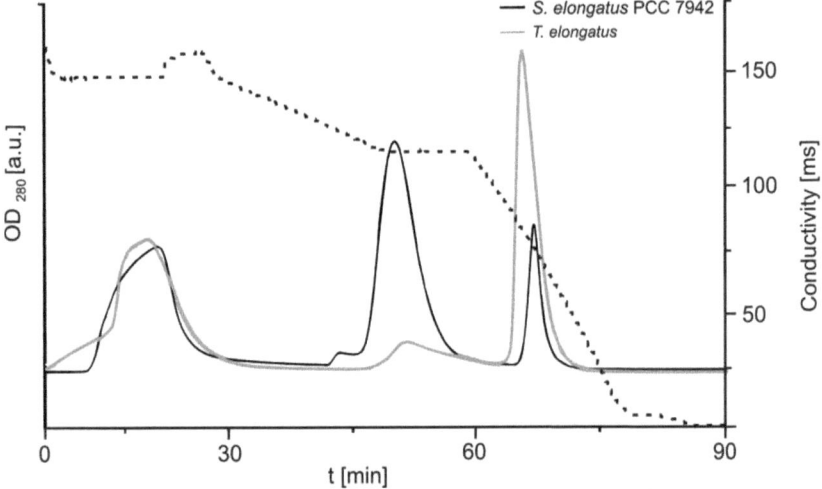

Figure 3.15: Elution profile of solubilised thylakoid membranes from S. elongatus PCC 7942 WT. HIC was used as first purification step compared to an elution profile obtained by preparation of PS I complexes from T. elongatus (El-Mohsnawy 2007). The corresponding conductivity is shown as the dotted line. Thylakoid membranes were solubilised with 0.5% β-DM. After loading, a linear gradient of $(NH_4)_2SO_4$ (1.5-0 M) was used for elution and flow rate 5 mL/min. HIC elution profile of solubilised thylakoid membranes prepared as described above from S. elongatus PCC 7942 WT, since two different profiles are illustrated together, the relative optical density at 280 nm is given. The flow rate during elution was set to 5 mL/min.

3.3.4.2 Second purification step using IEC

The concentrated and desalted PS I-containing fraction obtained by HIC was loaded onto an equilibrated POROS HQ/M column (Applied Biosystems). After a washing step with dialysis buffer, PS I particles were purified performing a $MgSO_4$ gradient from 0 to 0.2 M. The HPLC program for the second chromatography step is given in the supplementary files. The resulting elution profile illustrated in Figure 3.16 was similar to a corresponding elution profile obtained by purification of PS I complexes from T. elongatus via IEC as second chromatography step (El-Mohsnawy 2007). Since the elution was performed using a decreasing salt gradient, the phycobilisome-containing fraction eluted at first, followed by the purified PS I particles illustrated by the second peak. The small third peak refers to PS II complexes, which were eluted under high salt conditions.

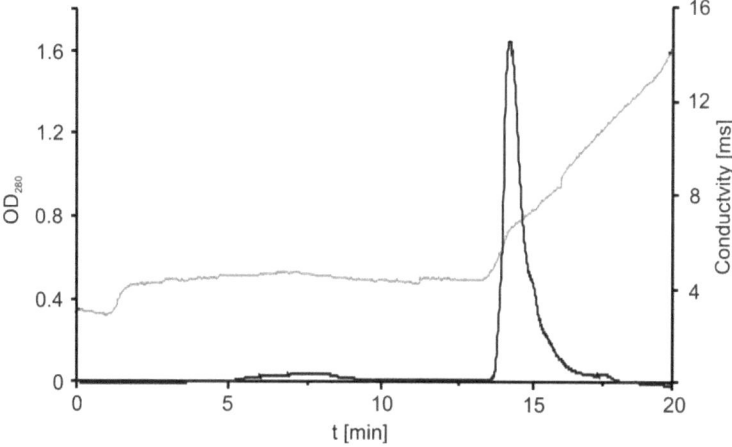

Figure 3.16: Elution profile of the *S. elongatus* PCC 7942 WT PS I fraction obtained by HIC as first purification step and using IEC (POROS 50HQ/M) as second purification step. Elution was performed by MgSO$_4$ gradient (0-0.2 M). Before elution, PS I was concentrated and dialysed against dialysis buffer overnight. OD$_{280}$ is given in black; the grey line shows the corresponding conductivity in ms. Flow rate during elution was set to 3 mL/min.

All obtained IEC fractions were concentrated by centrifugation using Centricon™ (Milipore) concentrators and desalted over night by dialysis against dialysis buffer. The concentrated samples were subsequently investigated by 77K measurements, immunoblot analyses, and BN PAGE. 77 K measurements showed that the PS II fraction, obtained by HIC, contained no detectable PS I particles (see Figure 3.17).

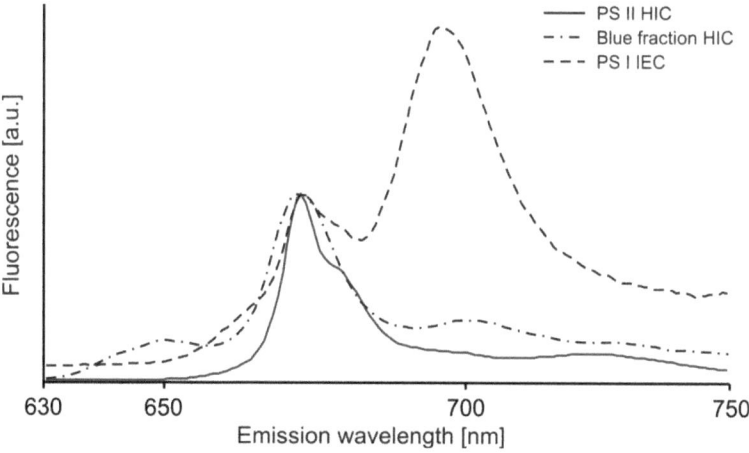

Figure 3.17: 77K pigment fluorescence emission spectra of selected fractions obtained by PS I purification with first HIC step and a second IEC step. The PS II fraction as well as the third phycobilisome-containing fraction obtained by HIC as well as the purified PS I fraction of the IEC are shown. Samples corresponding to 10 µg Chl *a* were measured. It became obvious that the PS I fraction, obtained by IEC, contained highly-purified PS I complexes, although a minor contamination with PS II particles existed.

On the contrary, the PS I fraction obtained after the second purification step using IEC showed in addition to a high content of PS I a significant peak at 695 nm and thus, included a detectable amount of PS II subunits. For the third HIC fraction, a significant phycobilisome content was detected by 77K measurements (685 nm peak) as well as high amounts of PS II and a minor amount of PS I. Immunoblot analyses of all obtained fractions (Figure 3.31 A) showed that the peak at 695 nm detected by 77K measurements in the phycobilisome-containing HIC fraction as well as the PS I IEC fraction must be caused by the integral light harvesting subunit CP47, because the main subunit of PS II, D1, was not detected in these fractions. Thus, these fractions contained only scattered subunits of PS I. Additionally, PsaAB was detected only in minor amounts in the third HIC fraction. In general, the results obtained by 77K measurements were confirmed by immunoblot analyses performed with the anti-PsaAB and the anti-D1 antiserum. The PS I IEC fraction contained high amounts of PsaAB as well as the PS II fractions of both chromatography steps contained high amounts of the D1 protein. The main content of PS II was removed during the first purification step by HIC.

Finally, the IdiC protein was not detected in any of the obtained fractions. As illustrated in Figure 3.18 A, only the phycobilisome-containing HIC fraction showed a faint protein band detected by the anti-IdiC antiserum. In contrast to the usually detected mass of IdiC corresponding to 23 kDa, the size of the detected band corresponded to 60 kDa indicating either an artefact (which is very unlikely) or a possible attachment of IdiC to other polypeptides. Results obtained by BN PAGE provided evidence that the previous solubilisation using 0.5% β-DM could be a reason for the loss of IdiC. Neither the PS I-containing IEC fraction, nor the PS II-containing fractions contained functional PS I trimers or PS II dimers, just monomers were shown (see Figure 3.18 B). Only the PS I fraction obtained by IEC as well as the phycobilin-containing HIC fraction contained significant amounts of PS I dimers. Possibly the used detergent concentration was high enough to disassemble PS I trimers and PS II dimers to monomers and peripheral subunits like CP47 of PS II or IdiC of PS I were completely separated and in the case of IdiC removed from the column during washing steps or elution.

To summarise the results obtained from investigations concerning the localisation of the IdiC protein in *S. elongatus* PCC 7942, it has to be stated that IdiC is loosely attached to the thylakoid membranes via connections to polypeptides of unidentified multi-protein complexes. The immunocytochemical investigations and immunoblot analyses with intact cells and subcellular fractions of *S. elongatus* PCC 7942 WT, respectively, as well as the anti-IdiC antiserum detected a large part of the cellular IdiC pool in the thylakoid membranes. Any evidence for a location of IdiC either in the cytoplasmic protein fraction or in the soluble protein fraction of the periplasm was not found yet.

Immunoblots with different antisera and gelelectrophoretically resolved multi-protein-complexes via two-dimensional PAGE indicated that parts of the cellular IdiC pool were associated with PS I complexes. These results were not confirmed by purification of PS I complexes from *S. elongatus* PCC 7942 WT thylakoids using HIC and IEC.

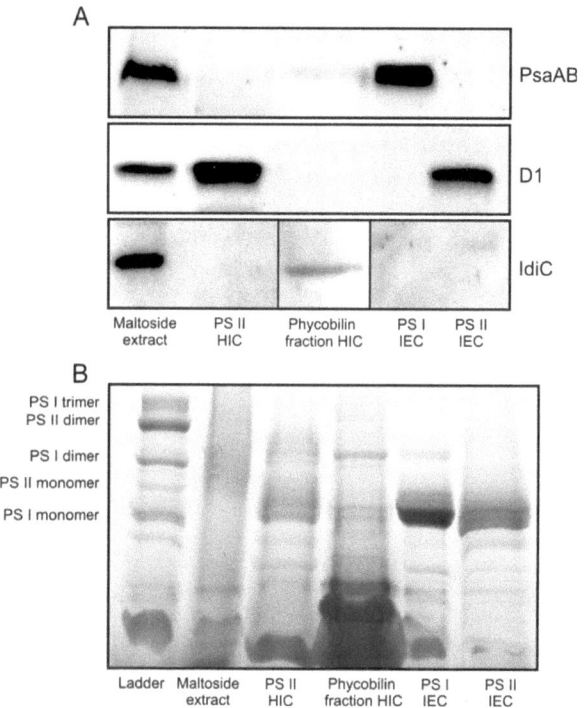

Figure 3.18: Investigation of protein fractions obtained from PS I purification from solubilised thylakoid membranes of *S. elongatus* PCC 7942 using a HIC and an IEC step. **(A)** Immunoblot with the anti-PsaAB (dilution 1:1,000), the anti-D1 (dilution 1:2,000), and the anti-IdiC antiserum (dilution 1:300). 100 µg protein of each fraction were subjected to SDS PAGE and immunoblotting. The IdiC protein was not detectable in any of the fractions, except a faint band in the 60 kDa range, which was detected in the phycobilin-containing third HIC fraction. **(B)** BN PAGE of the soluble protein fractions applying protein amounts corresponding to 50 µg protein. The PS I and PS II fractions obtained by IEC were highly enriched in PS I and PS II monomers, respectively.

3.4 Identification of the iron cofactor of IdiC

Since the bioinformatic analyses provided evidence that IdiC belongs to the family of TRX-fold [2Fe-2S] ferredoxins, an important aim of my work was to identify the putative iron-containing cofactor of IdiC. Therefore, the 6 x His-tagged IdiC protein was heterologously expressed in *E. coli*. However, all attempts to express soluble IdiC protein in elevated amounts always resulted in the formation of insoluble protein. Thus, 6 x His-tagged IdiC protein was homologously overexpressed in *S. elongatus* PCC 7942 mutants. Since the expressed IdiC amounts were low, and the protein formed insoluble aggregates after preparation of cell-free extracts, the heterologous expressed IdiC protein was detergent-solubilised and used for all subsequent investigations on the cofactor of IdiC.

3.4.1 Homologous expression of 6 x His tagged IdiC in *S. elongatus* PCC 7942

For homologous expression of IdiC in *S. elongatus* PCC 7942, the *his6idiC* nucleotide sequence of pQE82L*idiC* was cloned in p39, a genome-integrating vector for use in

S. elongatus PCC 7942. Plasmid p39 is a derivative of the plasmids pAM2318 and pAM2255 (see Table 2.3) and contained a spectinomycin resistance cassette. For insertion of the target sequence into the chromosome via homologue recombination, it contains the neutral site 1 (NS1) of pAM2255. The expression of the target protein is under the control of the *lac* system as known from *E. coli*. In the absence of IPTG, the *lac* repressor binds directly between the Ptrc promotor and thus, preventing transcription of the target gene. The *his6idiC* sequence was cloned into a single *Nco*I site, and the corresponding construct p39*his6idiC* was subsequently transformed in *S. elongatus* PCC 7942 WT. The resulting clones were selected for spectinomycin resistance and cultivated in BG11 medium containing 20 µg/mL spectinomycin. After verification of putative clones by colony PCR (see Figure 3.19 A), three clones were chosen for further investigations.

For overexpression studies of His6IdiC in *S. elongatus* PCC 7942, the mutants #11, #18, and #23 were cultivated in regular BG11 medium with spectinomycin. After 72 h of growth, the expression of IdiC was induced by the addition of 2 mM IPTG, and the appearance of the cultures as well as the corresponding growth was recorded. Figure 3.19 B shows that the impact of the IdiC overexpression caused a dramatically changed phenotypical appearance. Within 24 h, the appearance of the IdiC-expressing cell cultures changed from green to deep yellow, and the mutant showed no more substantial growth at all (data not shown) indicating a severe stress situation. It became obvious that clone #18 showed no bleaching after 24 h of expression. Subsequent investigations showed that this mutant was impaired in expression of the His-tagged IdiC protein and therefore, was used as a negative control (see Figure 3.21 B). Nevertheless, the phenotype of the mutants #11 and #23 resembled an appearance of *S. elongatus* PCC 7942 WT during prolonged growth under iron-deficient conditions, even though the bleaching occurred already during expression for 24 h.

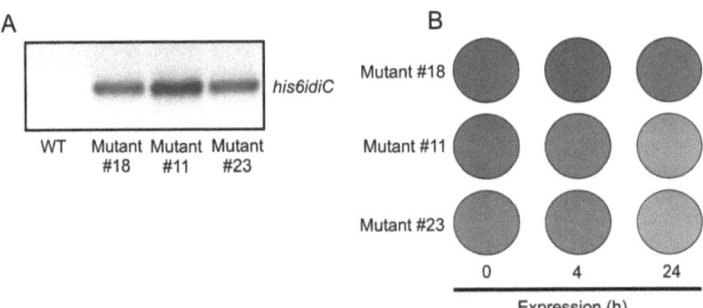

Figure 3.19: Construction and growth of IdiC overexpression mutants. The vector p39*his6idiC* was transformed in *S. elongatus* PCC 7942 WT introducing the *his6idiC* sequence into the bacterial genome via homologous recombination. The mutants #11, #18, and #23 were chosen for further investigations. **(A)** Detection of the *his6idiC* nucleotide sequence by colony PCR using gene-specific primers. **(B)** Appearance of cell cultures before and during the expression of His6IdiC induced by addition of 2 mM IPTG.

To investigate whether the bleached phenotype of the IdiC overexpression mutants during IdiC expression was based on major reduction of iron in the medium, cultures of mutant #11 were grown for 72 h in regular BG11 medium and treated with 30 µM supplemental FeCitrate and NH₄FeCitrate. The cultures were grown for an additional hour and the IdiC expression was initiated by adding IPTG to give a final concentration of 2 mM. Cell appearance and expression of IdiC was monitored after 14 h expression. Figure 3.20 A shows that cells were significantly bleached after an expression of IdiC for 14 h, when the medium was not treated with supplemental iron. The corresponding extent of His6IdiC expression was rather low as illustrated in Figure 3.20 B. In contrast, cultures, which received supplemental FeCitrate, showed no bleached phenotype and a high amount of overexpressed IdiC after 14 h. Interestingly, cultures treated with NH₄FeCitrate showed the highest degree of pigment degradation and simultaneously a low IdiC expression level comparable to the control. Thus, it could be stated that the phenotype of the mutants was dependent on the iron status of the culture and the culture medium. During expression of the iron-containing IdiC protein, supplemental iron led to an increased IdiC content and prevents the loss of photosynthetic pigments caused by iron deficiency. However, the addition of supplemental ammonium ions led to a decreased pigment content and physiological activity caused by an impaired assimilation of nitrate.

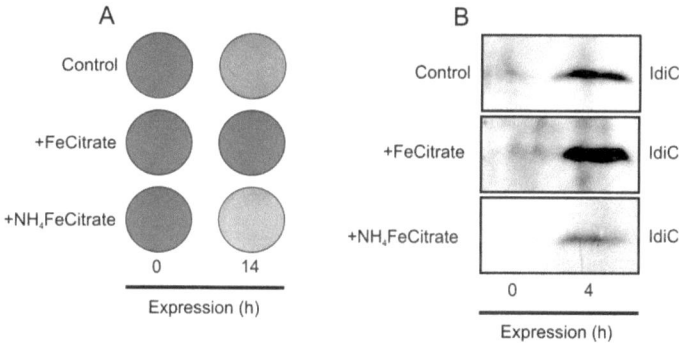

Figure 3.20: Phenotypical appearance of mutant #11 and the corresponding IdiC overexpression after treatment with supplemental iron. Cells were grown for 72 h in regular BG11 medium. After addition of 30 µM FeCitrate or NH₄FeCitrate, cells were grown continuously for 1 h before induction of IdiC expression by adding 2 mM IPTG. **(A)** Culture pigmentation before and after overexpression of IdiC for 14 h. **(B)** Detection of IdiC by immunoblotting with the anti-IdiC antiserum (dilution 1:500). 1 mL cell culture was harvested and resuspended in 50 µL ß-DM-containing denaturing buffer. After denaturing 15 µL of the cell suspension was subjected to SDS PAGE and blotting. Cell appearance and IdiC expression depended on the iron content in the medium. Supplemental ammonium ions led to additional pigment degradation and a decreased IdiC expression.

Further investigations illustrated in Figure 3.21 A showed that the iron-dependent phenotype of the IdiC overexpression mutants occurring during IdiC expression in regular BG11 medium was completely abolished by the addition of supplemental FeCitrate. Furthermore, the time course of the IdiC expression was examined for 24 h in the mutants #11 and #23. Despite the addition of 30 µM FeCitrate 1 h before initiation of the IdiC expression, the maximum content of IdiC was detected within the first 3 h (see Figure 3.21 B). These results suggested again an iron dependence of IdiC expression and implied an iron-containing cofactor of IdiC.

This suggestion was supported by the phenotype of mutant #23, which showed in general a lower pigment content under regular growth conditions as well as after addition of supplemental iron (see Figures 3.19 and 3.21). Immunoblot analyses shown in Figure 3.21 C provided evidence that this mutant possessed a significant IdiC overexpression without induction with IPTG and thus, had permanent iron deficiency already before induction of IdiC expression.

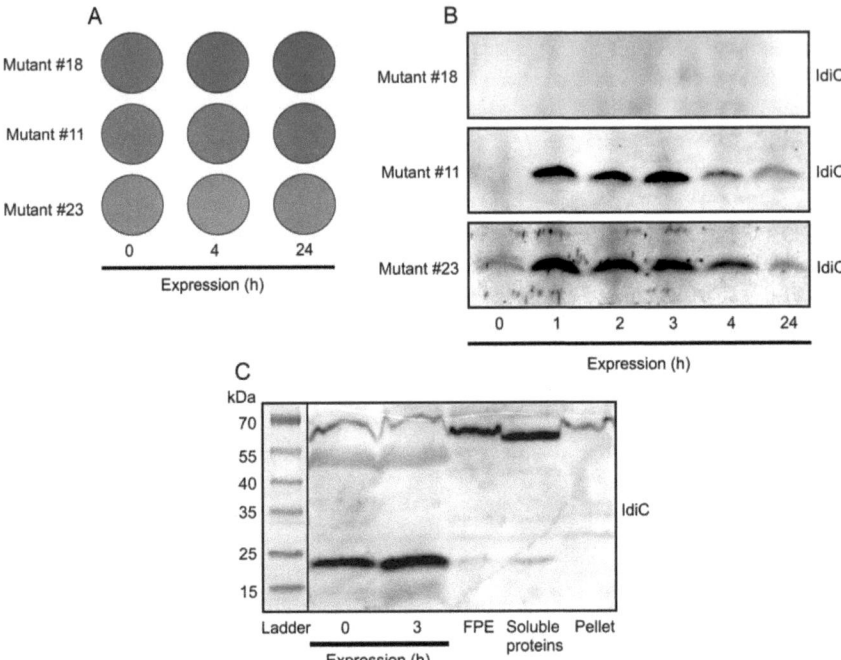

Figure 3.21: Time course of IdiC overexpression in *S. elongatus* PCC 7942 mutants #11 and #23 after addition of 30 µM supplemental FeCitrate. Cells were grown for 72 h under regular growth conditions and FeCitrate was added 1 h before induction of IdiC overexpression with 2 mM IPTG. **(A)** Appearance of cells before and during IPTG-induced expression of IdiC. **(B)** Time course of IdiC expression detected by immunoblotting with the anti-IdiC antiserum (dilution 1:300). 1 mL cell culture was harvested and resuspended in denaturing buffer. 15 µL of this cell suspension was subjected to SDS PAGE and immunoblotting. **(C)** Immunoblot analysis with the anti-IdiC antiserum (dilution 1:500) detecting IdiC after overexpression for 3 h in cells of mutant #23 and the corresponding cell-free extracts. Soluble proteins were extracted by centrifugation at 12,000 x g. Culture samples were treated as described above. Cell-free extracts corresponding to 50 µg protein were subjected to SDS PAGE and immunoblotting.

Since the amount of overexpressed His6IdiC protein was rather low and accumulated into soluble multimers during preparation of cell-free extracts, it was not used for further investigations about the identification of the cofactor (see Figure 3.21 C). However, the phenotype of the overexpression mutants during expression of IdiC under regular growth conditions compared to those occurring after addition of supplemental iron to the medium, provided evidence that the expression of IdiC severely affects intracellular iron homeostasis and thus suggest that IdiC contains indeed an iron-containing cofactor. Further investigations were carried out using heterologously expressed His6IdiC protein.

3.4.2 Heterologous expression of 6 x His-tagged IdiC in *E. coli*

To obtain high amounts of His-tagged IdiC protein for an investigation of the corresponding cofactor, the *idiC* nucleotide sequence of *S. elongatus* PCC 7942 was amplified via PCR with gene specific primers which include restriction sites for *Bam*HI and *Hind*III. After the restriction enzyme digest, the PCR product was cloned into the *Bam*HI and *Hind*III digested pET32a vector (Novagen), a plasmid designed for IPTG-inducible expression of N-terminal 6 x His-tagged proteins in *E. coli*. This expression system was chosen, because pET32a provides a C-terminal TRX tag to increase solubility of the heterologous expressed target protein. Previous heterologous expression of His6IdiC using the pQE expression system (Qiagen) produced almost insoluble IdiC protein (inclusion bodies). This protein was solubilised later on under denaturing conditions by the addition of 0.3% SDS and used to raise a polyclonal antibody against IdiC in rabbit (Pietsch 2004). Since the pET vector system included a T7 promoter, which is controlled by the LacI repressor, the pET32a*idiC* construct was transformed in *E. coli* strain BL21DE3, which possessed a chromosomal copy of the gene for T7 DNA polymerase. The resulting BL21DE3 clones carrying the correct plasmid were ampicillin resistant and were further verified by DNA sequencing.

Heterologous IdiC expression studies with the pET system were carried out under various conditions to optimise the yield and solubility of expressed IdiC. The growth temperature as well as the IPTG concentrations were altered to obtain substantial amounts of soluble His6IdiC protein. As a result, growth conditions at a temperature of 18 °C for 19 h were found to be optimal. Expression was induced with 0.2 mM IPTG at a cell density corresponding to OD 600 nm of 0.5 to 0.7. Cells were cultivated during expression in LBG medium containing 100 µg/mL ampicillin as well as 10 µM FeCitrate. After cell harvest, the corresponding cell-free extracts were prepared with a French Press treatment at 20,000 Psi. Soluble proteins were separated from the insoluble fraction via centrifugation at 12,000 x g. The IdiC content of the obtained protein fractions was controlled using SDS PAGE.

The results illustrated in Figure 3.22 A show that the His6IdiC content of cells harvested after 19 h overexpression were extremely high. However, the high content of the overexpressed protein led to an accumulation of insoluble protein and thus, the major His6IdiC amount was found in the pellet after centrifugation of the corresponding cell-free extracts. Image analysis provided evidence that the His6IdiC content was about 80% (ImageQuant™ software, Molecular Dynamics) of the entire cellular protein. Interestingly, the cell-free extract as well as the corresponding insoluble protein fraction showed a brownish colour in contrast to colourless protein fraction from cells, which carried an empty pET32a vector (see Figure 3.22 B and C). These cells served as a control in the following investigations. Due to the high amount of overexpressed His6IdiC in the samples and the fact that IdiC has similarity to [2Fe-2S] ferredoxins, it was highly likely that the brownish colour was due to iron bound to IdiC.

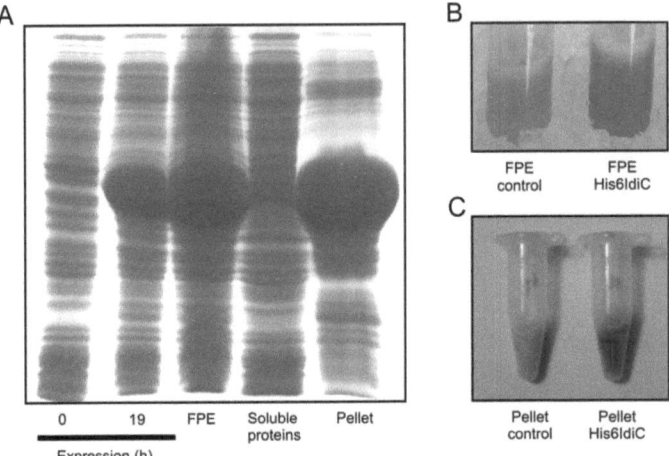

Figure 3.22: Heterologous expression of His6IdiC using the pET system from Novagen in *E. coli* strain BL21 for 19 h at 18°C. Cells were cultivated in LBG medium to which 10 µM FeCitrate was added. His6IdiC expression was induced by addition of 0.2 mM IPTG. **(A)** Coomassie-Brilliant-Blue-stained SDS PAGE. 1 mL samples of cultures were pelleted and resuspended in 50 µL denaturing buffer. Samples were applied to SDS PAGE corresponding to the appropriate OD_{600} value. **(B)** Comparison of cell-free extracts of a control, which carries the empty pET32a vector and the overexpression *E. coli* strain according to (A). **(C)** Comparison of the FPE pellets after preparation of insoluble proteins.

3.4.3 ICP-OES-based determination of metal content

For overall determination of metal ions via inductively coupled plasma spectroscopy, His6IdiC overexpression approaches were performed as described in chapter 3.4.2. The His6IdiC-containing insoluble fraction obtained from the corresponding cell-free extracts were transferred to 1.5 mL test tubes and dried at 60°C using a Speed Vac (Savant). After determination of the corresponding dry weight, samples were dry ashed at 480°C for 8 h and dissolved in 6 M HCl with 1.5% (w/v) hydroxylammonium chloride. Subsequently, this solution was diluted 1:10 with H2O bidest and used for determination of the concentration of metal ions via ICP-OES. The measurements were carried out in the lab of Prof. Dr. Walter J. Horst, Chair for Plant Nutrition at the University of Hannover. The results are presented in Table 3.2 as corresponding average values of 20 independent measurements. The major outcome of these experiments was that the IdiC-containing samples showed a two fold higher content of iron than the control and thus, suggested the presence of an iron-binding ability for IdiC. Since the side chains of the histidine residues forming the 6 x His tag are able to bind Ni+ as well as Zn+ ions, the His6IdiC-containing sample showed a six fold increased amount of Zn compared to the control. Thus, the results obtained from measurements of the Zn concentration, served as a good control for the significance of all presented ICP measurements. In contrast, the concentration of Mg, Al, and Cu, measured for the IdiC-containing sample, showed no alteration as compared to the control. In summary, the determination of metal ion concentration provided substantial data which support the presence of an iron cofactor within IdiC.

Table 3.2: Concentrations of selected metal ion determined by ICP-OES measurements for His6IdiC-containing insoluble French press fractions and the pET32a-empty vector control. Overexpression approaches were performed as described in chapter 3.4.2. The corresponding FPE pellets were dried, ashed at 480 °C, and subsequently dissolved in 8 M HCl. Compared to the control, the IdiC-containing samples possessed a two fold higher content of iron (yellow box). Since the Zn concentration measured for His6IdiC containing sample was six fold higher it served as a good control for the significance of these measurements (green boxed). DW = dry weight.

Element	Fe	Zn	Mg	Al	Cu
Sample	mg/g DW	mg/g DW	mg/g DW	mg/g DW	mg/g DW
Control	0.1089	0.0274	0.8397	0.0139	0.0042
His6IdiC	0.2309	0.1507	0.8868	0.0169	0.0039

3.4.4 Iron determination using EPR measurements

ICP-OES measurements suggested that IdiC contains an iron cofactor. EPR spectroscopy was performed to further investigate the nature of this cofactor. Since iron determination via EPR measurements requires a highly concentrated amount of soluble IdiC, solubilisation experiments were optimized first. The overexpression approaches were done as described in chapter 3.4.2. The resulting FPE pellet was resuspended in 10 mM HEPES pH 7.0. For solubilisation of the highly concentrated His6IdiC protein, β-DM was added to a final concentration of 2% (w/v), mixed, and samples were shaken gently for 1 h at 4 °C. Subsequently, samples were mixed in a potter and shaken again for another 30 min. The solubilised protein was separated by a centrifugation step at 12,000 x g. The results are illustrated in Figure 3.23. Almost the entire amount of His6IdiC was solubilised and the contamination with other proteins was found to be very low (see Figure 3.23 A).

A
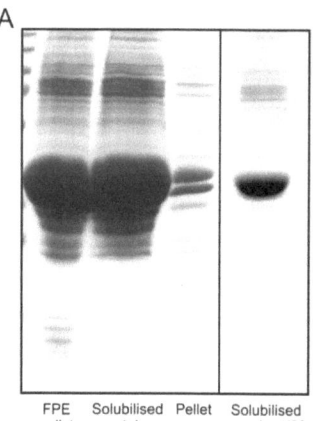

B
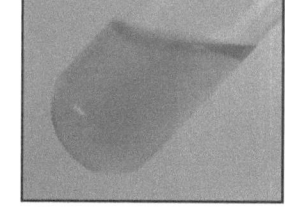

FPE pellet | Solubilised proteins | Pellet | Solubilised proteins 1/20

Figure 3.23: Solubilisation of the overexpressed His6IdiC protein with 2% (w/v) β-DM. Heterologous expression of IdiC was done as described in chapter 3.8.2. Cell-free extracts were prepared and the resulting insoluble protein fraction containing the His6IdiC protein was treated with β-DM and subsequently incubated at 4 °C for 90 min. Solubilised proteins were separated by centrifugation at 12,000 x g. **(A)** Coomassie Brilliant Blue-stained SDS PAGE to which 4 µL of each protein fraction was subjected. **(B)** His6IdiC-containing fraction with a concentration of 20.54 mg/mL after solubilisation.

Results

The solubilised His6IdiC protein was used for further investigations. At first, UV-Vis absorbance measurements in a range from 300 to 700 nm were performed. As illustrated in Figure 3.24 A, the IdiC-containing protein fraction showed a significant absorption peak around 420 nm being representative for bacterial ferredoxins. In detail, the spectrum of IdiC showed similarities to absorption spectra obtained for isolated bacterial ferredoxin II from *Clostridium thermoaceticum* (Elliott and Ljungdahl 1982) as well as to an oxidised peptide model of a [4Fe-4S] ferredoxin (Gibney et al. 1996). These results indicated that the iron bound by IdiC may indeed be arranged in a [Fe-S] cluster similar to that being found in ferredoxins as suggested by the bioinformatic analysis of the deduced amino acid sequence of IdiC (see Figure 3.24 B and C).

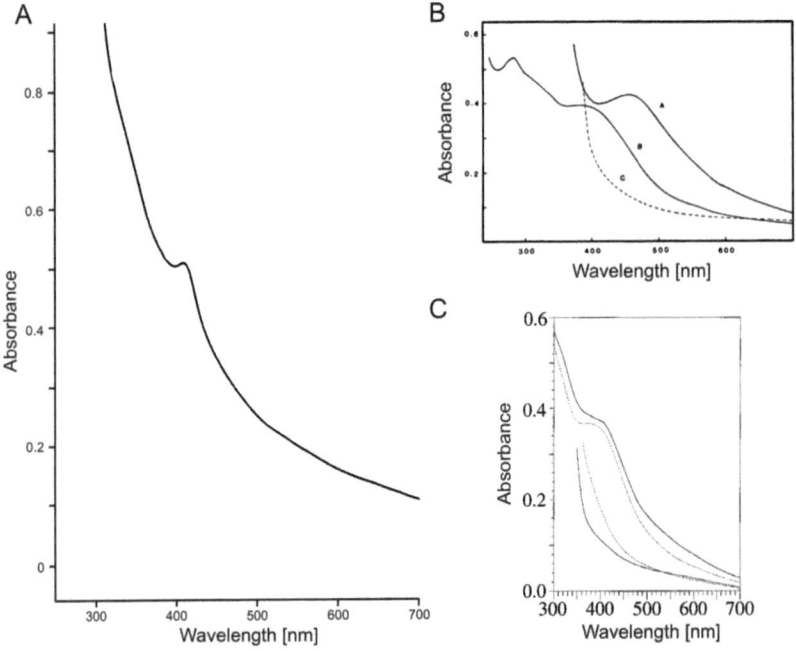

Figure 3.24: UV-Vis characterisation of the solubilised His6IdiC protein in 10 mM HEPES pH 7.0 compared to absorptions spectra of bacterial ferredoxins and a [4Fe-4S] ferredoxin maquette. Heterologous expression and solubilisation were performed as given in legend to Figure 3.36. **(A)** Absorption spectrum of solubilised His6IdiC (12.3 mg/mL). **(B)** Absorption of [Fe-S] centres separated from ferredoxin II of *Clostridium thermoaceticum* (A) as well as ferredoxin II, oxidised (B) and reduced (C), which was also isolated from *Clostridium thermoaceticum* (Elliott and Ljungdahl 1982). **(C)** Absorption spectra of oxidised and reduced [4Fe-4S] ferredoxin maquettes (Gibney et al. 1996).

In addition, these samples were used for cryogenic EPR spectroscopy. The solubilised His6IdiC samples were stored at 4 °C and shock frozen in liquid nitrogen. EPR measurements were carried out in cooperation with Dr. Edward Reijerse from the Max Planck Institute for Bioorganic Chemistry in Mülheim a. d. Ruhr with a Bruker EPR continuous wave X-band spectrometer E500 CW. A 50 mM Fe(III)EDTA pH 2.0 solution was used as control sample containing ferric iron. The resulting spectra are presented in Figure 3.25. Both spectra showed a significant rhombic signal at a high magnetic field range of

2,000 G. This result indicated that the His6IdiC-containg sample included a high content of ferric iron representing in an oxidised form. This signal allowed no final conclusions about the structure of the cofactor.

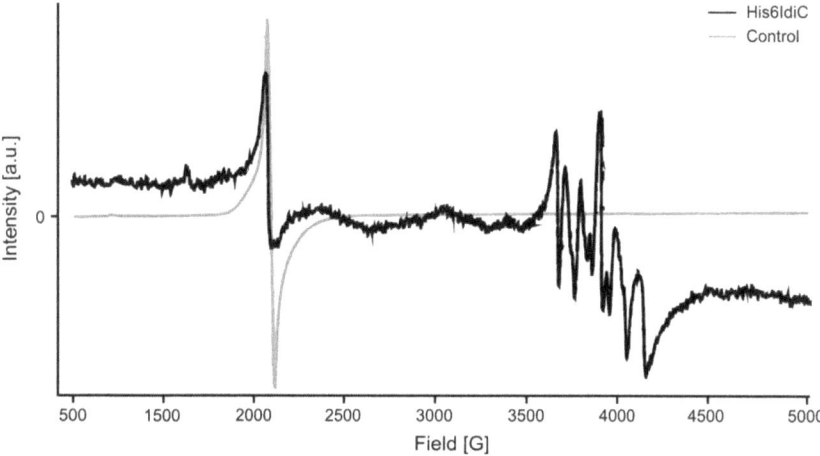

Figure 3.25: EPR spectra of the solubilised His6IdiC protein and a solution containing 50 mM Fe(III)EDTA, which served as a ferric iron containing control sample. Heterologous expression and solubilisation were performed as given in legend to Figure 3.36. 20.54 mg/mL protein was solubilised in 10 mM HEPES pH 7.0 containing 2% (w/v) β-DM. Conditions of EPR spectroscopy: Sample temperature 14K; microwave frequency 9.449 GHz; microwave power 0.2 mW; modulation frequency 100 kHz; modulation amplitude 20.0 G; scan field [G], 500 to 5000; averaged scans 42; and time constant 20.5 ms. The spectrum of the control is given in red.

In addition to the signal at high magnetic field, a couple of signals were detected in the range of 3,700 to 4,200 G. A similar signal at low magnetic field was reported to be characteristic for the oxidised form of bacterial [3Fe-4S]-containing ferredoxin (Aono et al. 1992) as well as oxidised [2Fe-2S] ferredoxin from Pseudomonas putida (Hugo et al. 1998). In addition, this signal was probably similar to analogous signals observed in preparations of oxidised bacterial ferredoxins, e.g. for that of Rhizobium japonicum (Carter et al. 1980). This signal was not assigned to a special [Fe-S] centre and may originate from the presence of superoxidised iron (Fe3+ oxidation level) as suggested by Sweeney at al. for the same type of signals observed in clostridial ferredoxins (Sweeney et al. 1974). Since all these spectra revealed only one rhombic signal, the line shape of the EPR signal shown for the His6IdiC-containing sample caused by the appearance of six different signals at low magnetic field was rather unusual for bacterial [Fe-S] clusters. Moreover, such a signal may have originated from Mn ions (Dr. E. Reijerse, MPI Mülheim, personal communication). However, the Mn content in the His6IdiC-containing sample was below the detection limit of ICP-OES measurements (data not shown) and thus, Mn can be almost entirely excluded to give rise to the detected signal. Moreover, it might be possible that this kind of signal was caused by the high concentration of detergents added to the sample or by conformational changes in the protein due to the solubilisation procedure. In addition, it cannot be entirely ruled out that the unusual signal shape of the His6IdiC-containing sample may as well be due to residual contaminating E. coli proteins, which were solubilised as well due to the addition of the detergent.

In conclusion, there are some indications that the solubilised IdiC protein may indeed represent a [Fe-S] ferredoxin - probably a somewhat atypical type of ferredoxin, which would fit the observation that it becomes expressed under conditions of iron starvation which is per se rather intriguing.

Since the bioinformatic analyses of the putative amino acid sequence showed strong similarities to the [Fe-S]-binding motif of known [2Fe-2S] ferredoxins, e.g. PetF2 from Synechococcus sp. PCC 7002, the presence of other [Fe-S] centre types in IdiC could almost be excluded. An alignment of the consensus binding motifs of the different so far known standard [Fe-S] clusters (Kaim and Schwederski 1995) with the corresponding [2Fe-2S]-binding motifs of PetF2 (Pfam Motif Search) from Synechococcus sp. PCC 7002 and IdiC is given in Figure 3.26. This shows that the [2Fe-2S]-binding motifs suggested for PetF2 and IdiC showed a higher similarity to each other than to the binding motif for [2Fe-2S] centres given in the literature. This fact implies that some differences between bacterial and cyanobacterial [Fe-S] clusters exist and would also explain in part the unusual signal shape at low magnetic field shown in the EPR spectrum of the His6IdiC-containig sample. Moreover, it has to be mentioned here that the similarity of the IdiC binding motif to the known binding motifs for [4Fe-4S] and [3Fe-4S] clusters were rather low (see Figure 3.26).

[2Fe-2S] ➡ - Cys - X_4 - Cys - X_2 - Cys - X_{29} - Cys -
PetF2 ➡ - Cys - X_4 - Cys - X_{26} - Cys - X_3 - Cys -
IdiC ➡ - Cys - X_4 - Cys - X_{27} - Cys - X_3 - Cys -
[4Fe-4S] ➡ - Cys - X_2 - Cys - X_2 - Cys - X_5 - Cys -
[3Fe-4S] ➡ - Cys - X_2 - Cys - X_n - Cys

Figure 3.26: Alignment of amino acid binding motifs for iron sulphur [Fe-S] centres (Kaim and Schwederski 1995) and the corresponding binding motifs suggested for PetF2 from *Synechococcus* sp. PCC 7002 and IdiC. It becomes obvious that the motif of IdiC showed higher similarities to that of PetF2 than to the consensus binding site of [2Fe-2S] centres given in the literature.

3.5 Construction and characterisation of an *idiC*- merodiploid *S. elongatus* PCC 7942 mutant

To elucidate the function of IdiC, a non-replicating vector construct for the insertional inactivation of *idiC* in *S. elongatus* PCC 7942 was cloned. However, all efforts to insertionally inactivate the *idiC* allele resulted in the generation of *idiC*-merodiploid mutants. The mutant with the highest degree of segregation was chosen for subsequent physiological investigations.

3.5.1 Attempts to construct an IdiC-free *S. elongatus* PCC 7942 mutant

To generate an IdiC-free *S. elongatus* PCC 7942 mutant, WT cells were transformed with plasmid pBSC9. This plasmid is a derivative of pKPM24 (Michel et al. 1996) carrying the entire *idiC* gene region on a 5.8 kb HindIII fragment of genomic DNA. A spectinomycin resistance cassette was cloned into a single EagI site of *idiC* to interrupt the gene (see Figure 3.27). The transformation experiment was done three times independently to obtain insertionally inactivated fully segregated *idiC* mutants.

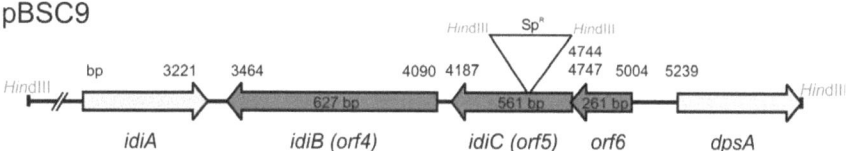

Figure 3.27: Partial physical map of plasmid pBSC9 used for construction of the *idiC* mutant. pBSC9 carries the entire 5.8 kb HindIII fragment of genomic DNA shown in database entry Z48754. The *idiC* gene was inactivated by insertion of a spectinomycin resistance cassette into an EagI site.

The obtained clones were spectinomycin resistant and sensitive to ampicillin as expected for a double cross-over transformation event. In order to achieve full segregation, clones were sub-cultured several times for two days with BG11 medium in the presence of increasing spectinomycin concentrations of up to 80 µg/mL.

To verify the insertional inactivation of the *idiC* allele in the obtained mutants, Southern blot analysis and colony PCR experiments were performed. Colony PCR assays with *idiC*-specific primers showed that all tested mutants still contained WT alleles next to insertionally inactivated *idiC* alleles (see Figure 3.28 A). The results were confirmed by Southern blot analysis with HindIII-digested genomic DNA obtained from *S. elongatus* PCC 7942 WT as well as the putative *idiC*-free mutants and a PCR-derived 3.5 kb Z48754-specific Dig-dUTP-labeled probe. The results of these analyses are shown in Figure 3.28 B. Thus, both experiments indicated that the mutants were *idiC* merodiploid, containing the WT alleles beside the insertionally inactivated mutant alleles.

For all subsequent experiments presented in this work, the clone with the highest degree of segregation was used and named MuD. Mutant MuD contains about 80% of insertionally-inactivated *idiC* alleles and about 20% WT *idiC* alleles (see Figure 3.28 B). The decrease of *idiC* WT alleles in mutant MuD resulted in a highly decreased amount of IdiC protein as compared to WT. An immunoblot experiment with the anti-IdiC antiserum and cell-free extracts from *S. elongatus* PCC 7942 WT and mutant cells grown for 96 h with BG11 medium is shown in Figure 3.28 C. 100 µg protein were applied to SDS PAGE and immunoblotting. Detection with the anti-IdiC antibody was done with the ECL™ detection kit (GE Healthcare). The mutant contained very low amounts of IdiC. Minor amounts of the protein were detected, when a sensitive detection method like ECL™ detection was used. Thus, the inability to construct a fully-segregated IdiC-free mutant suggests that IdiC is essential for the viability of *S. elongatus* PCC 7942 under the given experimental growth conditions.

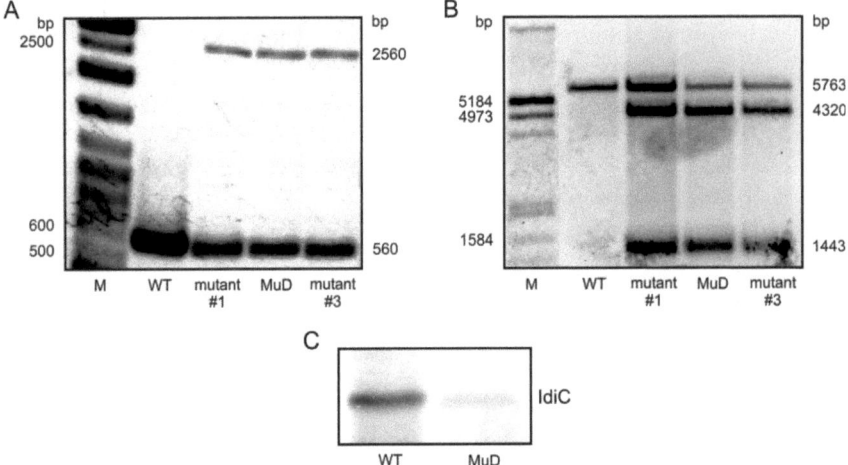

Figure 3.28: Verification of the insertional inactivation of *idiC* in *S. elongatus* PCC 7942 **(A)** Colony PCR of WT and three *idiC S. elongatus* PCC 7942 mutants with *idiC*-specific primers. The mutants showed amplified products for the *idiC* inactivated allele (2,560 bp) as well as for the *idiC* WT allele (250 bp). **(B)** Southern blot analysis with a 3.5 kb Z48754-specific Dig-dUTP labelled DNA probe and *Hind*III digested genomic DNA from WT and mutant MuD. The results showed that mutants were merodiploid and contained *idiC* WT alleles (5,763 bp fragment) as well as insertionally-inactivated *idiC* alleles (4,320 and 1,443 bp fragments). **(C)** Immunoblot with the anti-IdiC antiserum (dilution 1:300) and cell-free extracts from WT and mutant grown for 96 h under iron-sufficient conditions. Cell-free extracts corresponding to 100 µg protein were subjected to SDS PAGE and immunoblotting. The *idiC*-merodiploid mutant MuD contained minor, but detectable amounts of IdiC when immunostaining was done with the chemiluminescent ECL™ detection kit (GE Healthcare).

3.5.2 Effect of a reduced IdiC content on growth and photosynthetic activity of *S. elongatus* PCC 7942 WT and the *idiC*-merodiploid mutant MuD

For investigation of physiological effects caused by a reduced cellular IdiC amount growth, pigment content, and photosynthetic activities were determined for *S. elongatus* PCC 7942 WT and the *idiC*-merodiploid mutant MuD. Cells were grown with iron-sufficient and iron-deficient BG11 medium for 48, 72, and 96 h. Inoculation of cell cultures was performed with an optical density at 750 nm of 0.4.

Growth rates of mutant MuD under iron-sufficient conditions were similar to those of WT cells. When cells were grown under iron-deficient conditions, growth of MuD was significantly lower as compared to that of WT. The corresponding growth curves are given in Figure 3.29 A. Under regular growth conditions, the appearance of mutant MuD was almost comparable to that of WT cells. However, iron limitation led to decreased pigment content in

MuD and after growth of 96 h iron-starved mutant cells were highly bleached. The colour of the corresponding iron-deficient cultures turned from dark green into a deep yellow. In contrast, iron starved WT cells kept their green-blue colour even after 96 h under iron-deficient growth (see Figure 3.29 B). For detection of culture appearance 0.3 mL cell suspension aliquots were scanned at the times indicated using a HP Scanjet 7400c scanner.

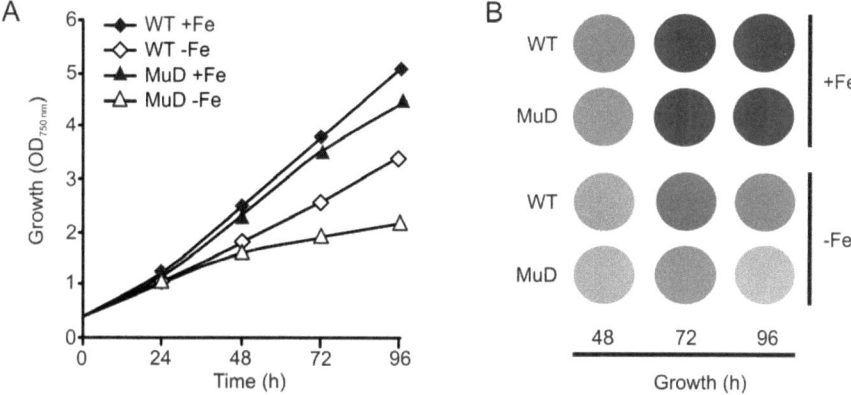

Figure 3.29: Growth curves and appearance of *S. elongatus* PCC 7942 WT and mutant MuD when grown under iron-sufficient or iron-deficient conditions. **(A)** *S. elongatus* PCC 7942 WT and mutant MuD were grown either with iron-sufficient or iron-deficient BG11 medium for 96 h. ♦ WT in iron-sufficient BG11 medium; ◊ WT in iron-deficient BG11 medium; ▲ mutant MuD in iron-sufficient BG11 medium; Δ mutant MuD in iron-deficient BG11 medium. **(B)** Appearance of WT and mutant MuD cells from the experiments as shown under (A). Under iron-deficient growth conditions, cells of mutant MuD had a much lower pigment content than WT cells. After 96 h the cells showed a yellow colour indicating strong bleaching of Chl. Under regular growth conditions, pigmentation of MuD was found equal to that of WT (Pietsch et al. 2007).

The appearance of cells displayed in Figure 3.29 B showed that mutant MuD contained less phycobili proteins than *S. elongatus* PCC 7942 WT, especially under conditions of prolonged iron limitation. Table 3.3 shows that due to the appearance of the corresponding cultures, cells of WT and MuD contained after 96 h of growth under iron-sufficient conditions almost identical contents of phycocyanin (PC). The content of the second phycobilin in *S. elongatus* PCC 7942, allophycocyanin (APC), was slightly increased even under regular growth conditions. After growth for 96 h with iron-depleted BG11 medium, MuD revealed a substantially reduced content of APC and PC as compared to that of WT. Especially, the content of PC decreased about 95%, while the APC content was reduced about 62%. In iron-starved WT cells, the PC content was reduced about 78%, while the APC content was only slightly decreased. Thus, the distal phycobilisome light-harvesting antennas, which are loosely connected to the thylakoid membranes, were almost entirely degraded in mutant MuD in contrast to those of WT.

Table 3.3: Phycobilin content of *S. elongatus* PCC 7942 WT and mutant MuD. Cells were harvested after growth for 96 h under iron-sufficient and iron-deficient conditions. Iron-starved cells of mutant MuD contained less phycobilins than WT. Cell suspension (CS) corresponded to a concentrated cell suspension with an optical density at 750 nm of 100.

Strains	Medium	Phycocyanin µg/mL CS	Allophycocyanin µg/mL CS
WT	+ Fe	103	42
	- Fe	23	41
MuD	+ Fe	101	34
	- Fe	5	13

In addition to the phycobilin content, the Chl *a* content as well as the photosynthetic activity of *S. elongatus* PCC 7942 WT and mutant MuD were measured. Cells were cultivated in iron-sufficient or iron-deficient BG11 medium for 48, 72, and 96 h. Photosynthetic oxygen (O_2) evolution of intact cells was determined with sodium hydrogen carbonate ($NaHCO_3$) as electron acceptor and an illumination with red light of an intensity of 200 µmol photons/m^2 s. The O_2 evolving activity was calculated per mL cell suspension (CS) and per mg Chl. The unit "mL cell suspension" corresponds to a concentrated cell suspension with an optical density at 750 nm of 100. These results are representative of three independent experiments.

The phycobilin content, the Chl *a* content as well as the photosynthetic activity of iron-sufficient mutant MuD was almost comparable to those of WT in the early growth phase after growth for 48 and 72 h (see Table 3.4). In the later growth phase after 96 h of growth, the Chl *a* content and the photosynthetic activity were lower in MuD as compared to WT. The differences between WT and MuD with respect to Chl *a* content and photosynthetic activity were even higher after 96 h, when cells were grown under iron-deficient conditions. Under conditions of prolonged iron starvation, MuD revealed only 50% of the photosynthetic activity, which was measured for WT cells harvested under the same growth conditions.

Table 3.4: Growth, Chl a content, and photosynthetic O_2 evolution of *S. elongatus* PCC 7942 WT and mutant MuD grown with regular and iron-deficient BG11 medium for 48, 72, and 96 h. After 48 and 72 h Chl a content and photosynthetic activity determined for MuD, were comparable to that of WT. In the late growth phase after 96 h, growth of iron-starved MuD Cells revealed a significantly lower Chl a content as well as a lower photosynthetic O_2 evolution.

Strains	Medium	Growth	Chlorophyll a	Photosynthetic O_2 evolution	
				$H_2O \rightarrow NaHCO_3$	
				µmol O_2/	
		OD_{750nm}	mg/mL CS	mL CS x h	mg Chl x h
Growth for 48 h					
WT	+ Fe	2.59	0.33	127	382
	- Fe	1.87	0.32	72	226
MuD	+ Fe	2.60	0.34	135	382
	- Fe	1.86	0.24	58	243
Growth for 72 h					
WT	+ Fe	3.40	0.36	80	245
	- Fe	2.44	0.25	25	143
MuD	+ Fe	3.42	0.38	75	194
	- Fe	2.29	0.17	20	122
Growth for 96 h					
WT	+ Fe	4.79	0.42	62	147
	- Fe	3.03	0.22	23	91
MuD	+ Fe	4.78	0.33	40	119
	- Fe	2.64	0.15	9	55

For immunoblot investigations of selected proteins participating in electron transport reactions, cells were grown with iron-sufficient and iron-depleted BG11 medium for 48, 72, and 96 h. Cell-free extracts corresponding to 100 µg protein were subjected to SDS PAGE and immunoblotting. Detection was performed with antisera against PsbA, PsbO, PetA, PsaA/B, and FNR. Antisera were used at the dilutions given in chapter 2.34.5. Blots were developed using ECL™ detection kit (GE Healthcare).

Figure 3.30 shows that the lower photosynthetic activity was consistent with a lower content of the PS II proteins PsbA and PsbO in mutant MuD as compared to WT, especially after growth of 96 h with iron-deficient BG11 medium. Immunoblots with an anti-PsaA/B antiserum provided evidence that the amount of PS I reaction centre heterodimers in WT and MuD cells remained constant under iron-sufficient conditions within 96 h of cultivation. However, a strong decrease of the PsaA/B in WT and MuD was detected under initial as well as prolonged iron starvation. Detection with the anti-PetA antiserum showed that the amount of Cyt *f* in mutant MuD cells grown under iron-sufficient conditions was similar to that of WT, showing only a minor reduction in MuD with increasing culture age. The PetA content increased in both strains under iron-limiting growth conditions, supporting an increase in photosynthetic cyclic electron transport activity and/or respiration under these conditions.

Multiple forms of FNR have been suggested to be present in *Synechocystis* sp. PCC 6803 (Matthijs et al. 2002; Thomas et al. 2006; van Thor et al. 2000). However, so far no information about multiple FNR forms in *S. elongatus* PCC 7942 exists in the literature. Nevertheless, due to the results with *Synechocystis* sp. PCC 6803 with respect to FNR, in this work a peptide antiserum against the N-terminal part of FNR (anti-FNR1 antiserum) as well as against the C-terminal part of FNR (anti-FNR 9A antiserum) was raised to distinguish between the full-length protein and its processed forms. Under iron-sufficient growth conditions, there was no major difference between WT and mutant MuD with respect to FNR (see Figure 3.30). Almost the entire FNR (more than 90%) was detected in one band with an apparent molecular mass of 45 kDa. The calculated molecular mass of FNR is defined with 44.4 kDa. Under iron limitation the amount of FNR increased significantly in WT, but in contrast decreased significantly in MuD. In WT as well as MuD grown under iron limitation the FNR antisera recognized multiple FNR protein bands with different mobility. In WT two bands of approximately 45 and 35 kDa were detected by both antisera and therefore, represent the full-length protein, while the 30 kDa band was only recognized by the anti-FNR 9A antiserum raised against the C-terminal part and therefore, must represent a processed form (possibly without the CpcD-like domain, see Matthijs et al. 2002). In mutant MuD both antisera recognized the full-length protein of about 45 kDa, while a band of 36 kDa was only recognized by the anti-FNR 9A against the C-terminal part and thus, must represent a processed form. This processed form in MuD has a different electrophoretic mobility than the processed form in *S. elongatus* PCC 7942 WT. The difference in mobility of the two full-length FNR forms in WT and of the processed form in WT and MuD could possibly be due to an association with lipids and could indicate that these forms were originally present in a different environment in the thylakoid membrane and therefore, might have interacted with different lipids.

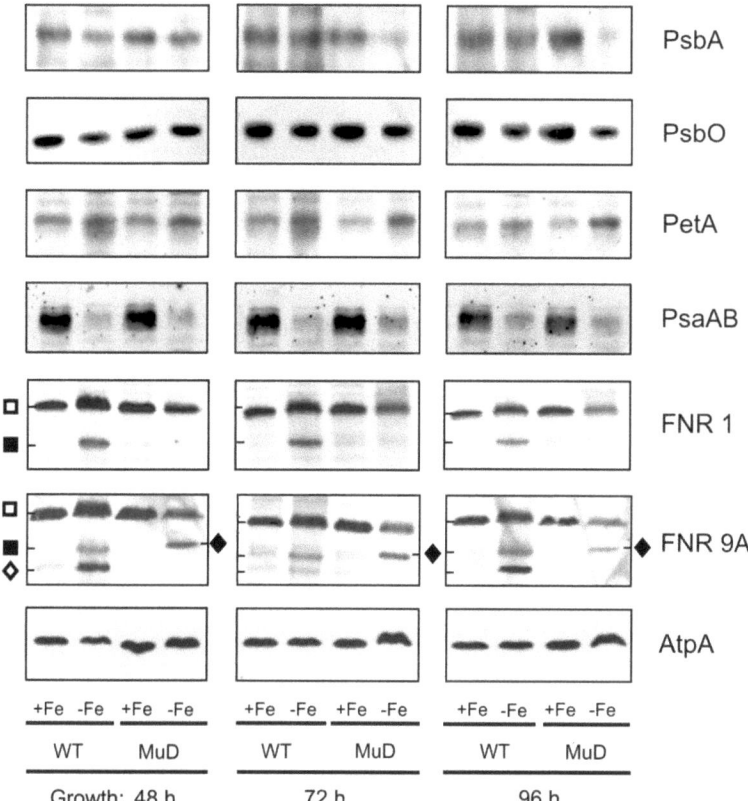

Figure 3.30: Immunoblot detecting selected proteins participating in electron transport reactions in MuD and *S. elongatus* PCC 7942 WT. Cell-free extracts corresponding to 100 µg protein were subjected to SDS PAGE and immunoblotting. Detection was performed with antisera against PsbA, PsbO, PetA, PsaA/B, and FNR. The applied dilutions of the corresponding antisera are given in chapter 2.34.5. The anti-FNR 1 antiserum was raised against an intrinsic peptide of the CpcD-like domain at the N-terminus of the FNR. The anti-FNR 9A antiserum was raised against an intrinsic peptide in the C-terminal region of FNR (see chapter 2.28). (□) full-length FNR and (■) modified full-length FNR (possibly associated with lipids resulting in a different electrophoretically mobility) detected with the anti-FNR 1 (N-terminal) and anti-FNR 9A (C-terminal) antisera; (◊) processed or N-terminally degraded FNR in WT and (♦) processed or N-terminally degraded FNR in MuD only detected with anti-FNR 9A (C-terminal) antiserum. Immunodetection with the anti-AtpA antiserum was included as an additional control to assure equal protein loading, since the AtpA protein content was not affected by iron starvation under our conditions (Exss-Sonne et al. 2000; Michel et al. 2003).

In summary, it can be concluded that mutant MuD exhibited a strong iron-dependent phenotype. Especially under prolonged iron starvation MuD showed a major reduced pigment content and photosynthetic activity compared to that of *S. elongatus* PCC 7942 WT. The reduced photosynthetic activity was due to a strong decrease of the main subunits of the photosystems, especially those of PS I. In addition, the amount of full-length FNR was increased under iron limitation in WT but decreased in mutant MuD. Whether the modified forms of the full-length FNR with an apparent molecular mass of 30 kDa are functional, is presently unclear.

3.5.3 Consequences of reduced IdiC content on the respiratory electron transport activity in the *idiC*-merodipolid *S. elongatus* PCC 7942 mutant MuD

As documented in previous chapters, iron limitation leads to a number of modifications of the photosynthetic and respiratory electron transport chain, which leads the way to a down-regulation of photosynthetic open-chain electron transport from water to NADP and to an up-regulation of photosynthetic cyclic electron transport around PS I as well as to an up-regulation of the respiratory electron transport activity. Since IdiC has similarity to NuoE, a substrate-binding subuntis of the NDH-1 complex in *E. coli*, it is likely that IdiC is a component of the NDH-1 complex or interacts with the NDH-1 complex in *S. elongatus* PCC 7942. In cyanobacteria the NDH-1 complex has a function in the cyclic electron flow around PS I and in respiration catalysing the oxidation of NADH with electron donation into the respiratory electron transport chain. Based on these considerations it was asked whether a reduced amount of IdiC in MuD has an effect on respiration.

To answer this question, the O_2 uptake of intact *S. elongatus* PCC 7942 WT and MuD cells was determined in an Clark type O_2 electrode under dark conditions. For these measurements, the cells were cultivated for 24, 72, and 120 h under standard conditions in the presence or in the absence of iron. After harvesting the cells, the cells were resuspended in the growth medium to give a cell density of 100 µL cell mL^{-1}. In the O_2 electrode, the endogenous O_2 uptake without addition of further substrates and the O_2 uptake in the presence of added L-arginine was determined. Moreover, the effect of $CaCl_2$, which inhibits the L-arginine oxidase/dehydrogenase and the effect of KCN, which inhibits the terminal oxidase of the respiratory electron transport chain and the catalase-peroxidase, were determined.

For a better understanding and for an interpretation of the obtained O_2 uptake values, a number of background information has to be taken into consideration. The cells can catalyse NADH oxidation via the NDH-1 complex feeding electrons into the respiratory electron transport chain. The O_2 uptake by the cytochrome oxidase is inhibited by KCN (see Figure 3.31 A). *S. elongatus* PCC 7942 contains an L-amino acid oxidase/dehydrogenase (L-Aox), which is expressed constitutively. The enzyme catalyses the oxidative deamination of basic L-amino acids with L-arginine being the best substrate utilising molecular O_2 as electron acceptor (O_2 uptake) leading to formation of keto-arginine, NH_4^+ and H_2O_2. This O_2 uptake is inhibited by $CaCl_2$. In the presence of catalase-peroxidase, being present in *S. elongatus* PCC 7942, the formed H_2O_2 becomes decomposed to ½ O_2 and H_2O. This O_2 evolution from H_2O_2 is inhibited by KCN (see Figure 3.31 C).

A number of results have indicated that part of the enzyme is associated with the respiratory electron transport chain. This suggests that electrons from L-arginine oxidation can be donated to the respiratory electron transport chain implying that the enzyme can also act as an L-arginine dehydrogenase. The O_2 uptake of this reaction is inhibited by $CaCl_2$, which inhibits the L-arginine dehydrogenase and by KCN, which inhibits the cytochrome oxidase (see Figure 3.31 B).

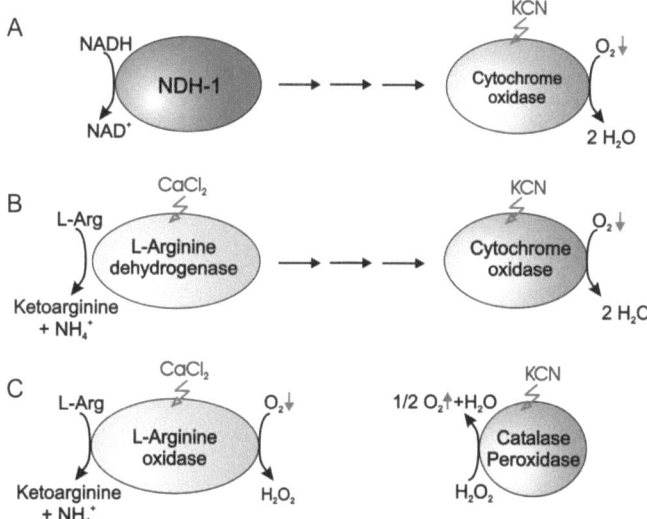

Figure 3.31: Schematic presentation of possible respiratory activities in *S. elongatus* PCC 7942 and the effect of inhibitors on these reactions. **(A)** NADH oxidation via the NDH-1 complex feeds electrons into the respiratory electron transport chain. The O_2 uptake mediated by the cytochrome oxidase is inhibited by KCN. **(B)** L-arginine oxidation via the L-arginine dehydrogenase feeds electrons into the respiratory electron transport chain. The O_2 uptake mediated by the cytochrome oxidase is inhibited by KCN. The L-arginine dehydrogenase activity is inhibited by $CaCl_2$. **(C)** L-arginine oxidation via the L-arginine oxidase catalyses the oxidative deamination of L-arginine to keto-arginine, NH_4^+, and H_2O_2-utilising molecular oxygen as electron acceptor. The L-arginine oxidase is inhibited by $CaCl_2$. In presence of catalase-peroxidase, which is present in *S. elongatus* PCC 7942, the H_2O_2 is decomposed to ½ O_2 and H_2O. This reaction is inhibited by KCN.

Since the L-arginine dehydrogenase reaction most likely does not proceed via the NDH-1 complex, it is likely that in the *idiC*-merodiploid mutant MuD with a reduced IdiC content a reduced regular respiratory is present, while the L-arginine dehydrogenase activity is increased. If this is the case, the L-arginine-stimulated and KCN-inhibited O_2 uptake should be found to be increased.

The results of these experiments are given below in Table 3.5. Since under iron limitation the Chl content per cell is reduced as compared to growth in presence of iron, the O_2 uptake is calculated on cell basis. The major results of these experiments are:

1. The O_2 uptake without added endogenous substrate is substantially lower in MuD than in WT after 24, 72, and 120 h of growth (see column A of Table 3.5).

2. Substantial variations can be seen in the L-arginine dehydrogenase-mediated respiratory activity (see column G of Table 3.5).

3. As expected, this O_2 uptake was always higher in MuD than in WT. Moreover, this L-arginine mediated O_2 uptake was higher in cells grown under iron limitation than in cells grown in iron sufficient medium.

Table 3.5: Measurement of the O_2 consumption, which was due to the L-arginine dehydrogenase catalysing the oxidative deamination with electron donation into the respiratory electron transport chain. For the calculation of the O_2 uptake that represents the L-arginine dehydrogenase mediated part, the facts given in the text are taken into consideration. The O_2 uptake values given in the column G represent that part of the O_2 uptake that is based on L-arginine oxidation and electron donation via the cytochrome oxidase to O_2 and thus, is a reaction that is inhibited by KCN. "Arg" Arginin.

Strains	Medium	O_2 uptake (µmol × mL CS^{-1} × h^{-2})						Calculated expected value	L-Arg respiration	L-Arg respiration % of total O_2 uptake
		Endog. substrate	Endog. substrate + added L-Arg	+ added L-Arg	Endog. substrate + KCN	Endog. substrate + added L-Arg + KCN				
		A	B	C (B-A)	D	E		F (B-C/2-(A-E))	G (F-E)	H G as % of B
Growth for 24 h										
WT	+ Fe	16.5	27.5	11.0	5.5	16.5		22.0	5.5	20
WT	- Fe	33.8	59.1	25.3	12.7	33.8		50.7	16.9	29
MuD	+ Fe	10.3	24.2	13.9	0.8	9.8		21.7	11.9	49
MuD	- Fe	12.8	24.2	13.9	0.8	9.8		21.7	9.3	35
Growth for 72 h										
WT	+ Fe	17.8	41.6	23.8	11.9	47.5		47.6	0.1	0
WT	- Fe	26.0	73.0	47.2	4.1	56.9		74.9	18.0	25
MuD	+ Fe	5.4	32.5	27.1	1.6	43.3		42.3	0	0
MuD	- Fe	7.4	47.3	39.9	0	35.4		59.9	24.5	52
Growth for 120 h										
WT	+ Fe	17.3	23.0	5.7	1.7	13.3		10.3	0	0
WT	- Fe	17.1	34.0	17.2	4.6	22.9		30.4	7.5	22
MuD	+ Fe	5.4	18.4	13.0	0	5.4		19.5	14.1	77
MuD	- Fe	5.9	14.7	8.8	2.2	7.3		15.4	8.1	55

In summary, it can be stated that the respiratory activity based on endogenous substrate is substantially lower in MuD than in WT. This is most likely due to a lower NDH-1 complex activity as a result of a reduced cellular IdiC amount.

In addition, the results show that the L-arginine dehydrogenase-mediated respiratory electron transport contributes to the overall cellular respiratory O_2 consumption. This contribution is higher in MuD than in WT. It is also higher in cells grown under iron limitation than in cells grown under iron-sufficient condition. This is the case in MuD as well as WT. The results suggest that in MuD with a lower IdiC content and thus, a lower NDH-1-activity the NDH-1-mediated respiration is reduced. To compensate for this deficiency, the L-arginine dehydrogenase-mediated respiratory activity is increased. Obviously, this activity represents a larger part of the overall respiratory activity in S. elongatus PCC 7942 under iron limitation.

3.5.4 Consequences of reduced IdiC content on the expression of selected iron-regulated genes and proteins in MuD compared to S. elongatus PCC 7942 WT

In S. elongatus PCC 7942 transcription of the *idiB* operon has been shown to be up-regulated under iron limitation by the action of a still unidentified transcription factor (Yousef et al. 2003). Furthermore bioinformatic investigations provided evidence that the IdiC protein might represent one of the so far unknown substrate-binding subunits of the cyanobacterial NDH-1 complex. Therefore, the consequences of the reduced IdiC protein level on the expression of a selected number of iron-regulated genes were investigated in detail in mutant MuD compared to S. elongatus PCC 7942 WT. Cells were inoculated at an optical density at 750 nm of 0.4, cultivated in iron-sufficient or iron-deficient BG11 medium, and harvested after 48, 72, and 96 h for isolation of total RNA and preparation of cell-free extracts.

At first, expressions of iron-regulated genes were analysed by Northern blot investigations. After extraction of total RNA, 10 µg RNA was subjected to gel electrophoresis and subsequently blotted to nylon membranes. Northern blots were performed with selected Dig-dUTP labelled probes (see Figure 3.32). Detection with an *idiC*-specific probe showed that the *idiC* message as well as the entire operon message increased in WT under iron limitation after 48 and 72 h. After 96 h of growth the message was already high under iron-sufficient conditions. This finding is consistent with results showing that the IdiC protein expression increased in the later growth phase (see Figure 3.7). In mutant MuD, an increase in the *idiC* and the entire *idiB* operon transcript level was detected after 48, 72, and 96 h of growth under iron limitation, although the amount of the mRNA was significantly lower in MuD than in WT. This is in agreement with the reduced number of WT alleles in mutant MuD as illustrated in Figure 3.28.

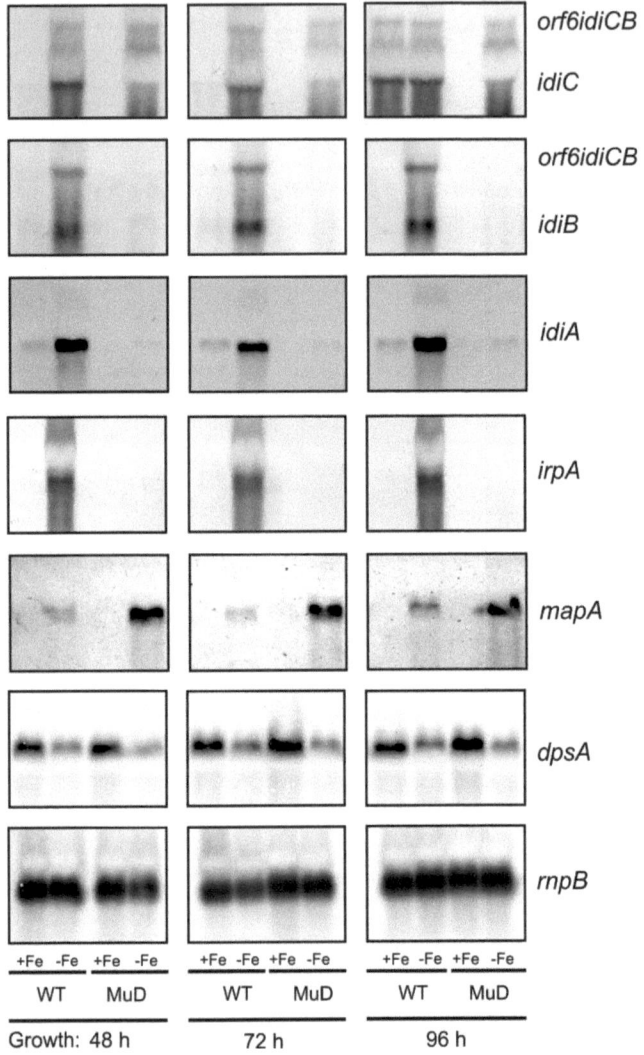

Figure 3.32: Analyses of selected transcripts of S. elongatus PCC 7942 WT and mutant MuD. 10 µg RNA was subjected to gel electrophoresis and subsequent blotted to nylon membranes. Dig dUTP-labeled DNA probes being specific for *idiC*, *idiB*, *idiA*, *irpA*, *mapA*, and *dpsA* were used to detect the corresponding steady-state transcript pools. An *rnpB*-specific probe was used to assure equal loading of RNA. The expression of IdiB and IdiB-regulated proteins was hardly detectable in mutant MuD.

In contrast to WT, the increase in *idiC* message in the later growth phase was not detected in mutant MuD. With an *idiB*-specific probe the increase of *idiB* and the entire operon message was clearly detectable in WT after 48, 72, and 96 h under iron starvation but not in mutant MuD. This was quite unexpected, since both genes, *idiB* and *idiC*, are located in the same operon. Another difference was that under iron-sufficient conditions no *idiB* mRNA was detected in WT in the late growth phase - in contrast to the detection of the *idiC* message.

This could be based on a higher stability of the *idiC* mRNA as compared to the *idiB* transcript. This might also reflect a higher demand for IdiC than for IdiB, which would be in agreement with the fact that IdiC is essential, while IdiB is not (Michel et al. 1999). A somewhat similar observation has been made for the *isiA/isiB* operon, being transcribed in a monocistronic and a dicistronic message (Straus 1994). The *isiA*-encoding monocistronic message has been found to be much more abundant than the dicistronic *isiAB* mRNA and might reflect a higher demand for IsiA than for IsiB (flavodoxin).

Since hardly any *idiB* mRNA is present in MuD under iron limitation, almost no *idiA* and *irpA* transcripts were detectable, because IdiB is the transcription factor for these two genes (Yousef et al. 2003). It has already been suggested that IrpA is located in an operon and that the genes of this operon encode proteins of an iron acquisition system (Reddy et al. 1988). Another major iron-regulated gene, *mapA* (Webb et al. 1994) was detected in higher amounts in MuD under iron starvation than in WT. MapA represents a membrane-associated protein of yet unknown function. No significant difference between WT and mutant MuD under iron-limiting growth conditions was detected for the expression of *dpsA* (Dwivedi et al. 1997; Marjorette et al. 1995), having a function in DNA protection under oxidative stress and in the stationary phase.

The protein expression pattern of IdiC, IdiB, IdiA, IsiA, IsiB, and DpsA was also determined by immunoblot analysis. Cells were grown as described above and harvested after growth for 48, 72, and 96 h with regular BG11 medium or BG11 medium from which iron was omitted. After cell harvesting, cell-free extracts were prepared and applied to SDS PAGE and blotting. Figure 3.33 shows that a strong increase of the IdiC protein was observed under iron starvation in WT, while in MuD IdiC was hardly detectable. This is consistent with the low number of *idiC* WT alleles in mutant MuD. Moreover, as already shown for the *idiC* mRNA level under iron-sufficient conditions, also the IdiC protein content increased during the later growth phase in WT. Thus, after 96 h of growth hardly any difference in the IdiC protein level of WT cells grown under iron-sufficient or iron-deficient conditions existed. The expression patterns of IdiB and DpsA in WT and MuD resembled the results obtained from Northern blot analyses. Since any IdiB expression was not detected in MuD, IdiA expression was also barely detectable in this mutant.

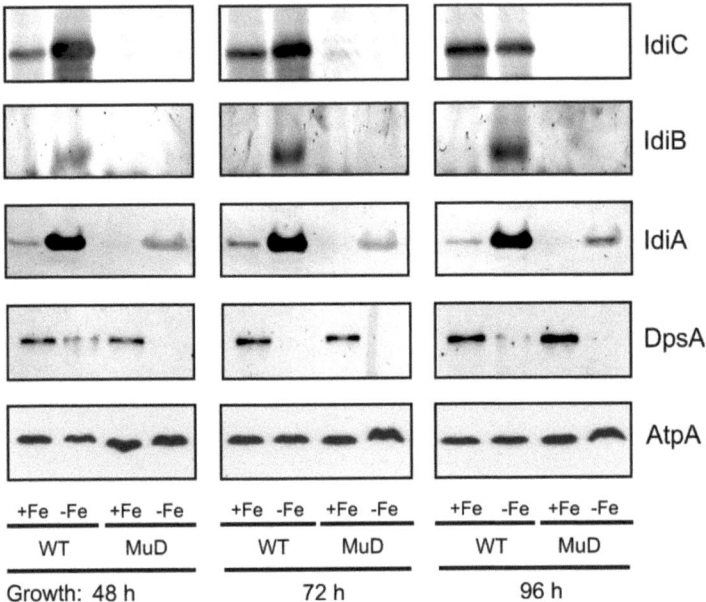

Figure 3.33: Immunoblots of selected iron-regulated proteins of S. elongatus PCC 7942 WT and mutant MuD. Cells were grown for 48, 72, and 96 h in regular and iron-depleted BG11 medium. Cell-free extracts corresponding to 100 µg protein were subjected to SDS PAGE and blotting in case of IdiC, IdiB, IsiA, and IsiB immunblots. In case of IdiA, DpsA, and AtpA blots, 50 µg protein were applied. Detection was performed with the anti-IdiA (dilution 1:2,000), anti-IdiB (dilution 1:200), anti-IdiC (dilution 1:300), anti-DpsA (dilution 1:1,000), anti-IsiA (dilution 1:500), anti-IsiB (dilution 1:1,000), and anti-AtpA antiserum (dilution 1:2,000). Immunodetection with the anti-AtpA antiserum was included to assure equal protein loading, since AtpA expression was not affected by iron starvation under the used conditions (Michel et al. 2003). The expression of IdiA, IdiB, IdiC, and DpsA was consistent with the findings of a Northern blot analysis.

For *isiA* and *isiB* no substantial differences between WT and mutant MuD under iron-limiting growth conditions were detected (see Figure 3.20 A). This result was anticipated, since the *isiAB* operon is not regulated by IdiB but by the transcriptional repressor Fur (Ghassemian and Straus 1996). Both genes showed an increased transcription rate under conditions of iron limitation. Whereas the amount of IdiA protein is in good agreement with the results of the Northern blot analysis (see Figure 3.32), the IsiA protein content did not as strongly increase as the *isiA* mRNA in iron-depleted MuD cells (see Figures 3.34 A and B). This deviation between *isiA* mRNA and protein level was not observed in WT and was also not observed for the *isiB* mRNA and the IsiB protein. Consistent with a low IsiA content, a low IsiA-related Chl *a* fluorescence peak at 685 nm was monitored in iron-starved MuD cells after a growth of 48, 72, and 96 h with iron-depleted BG11 medium (see Figure 3.34 C). This difference in *isiA* mRNA and protein level in mutant MuD suggests that a signal transduction pathway causing the up-regulation of IsiA protein expression is altered in mutant MuD. Thus, the absence of IdiC has an influence on either IsiA protein expression and/or IsiA protein degradation. It is tempting to speculate that IsiA expression is in part regulated by the redox status of components of the photosynthetic electron transport chain, e.g. the PQ pool. This implies a function of IdiC in photosynthetic electron transport under stress conditions like iron starvation.

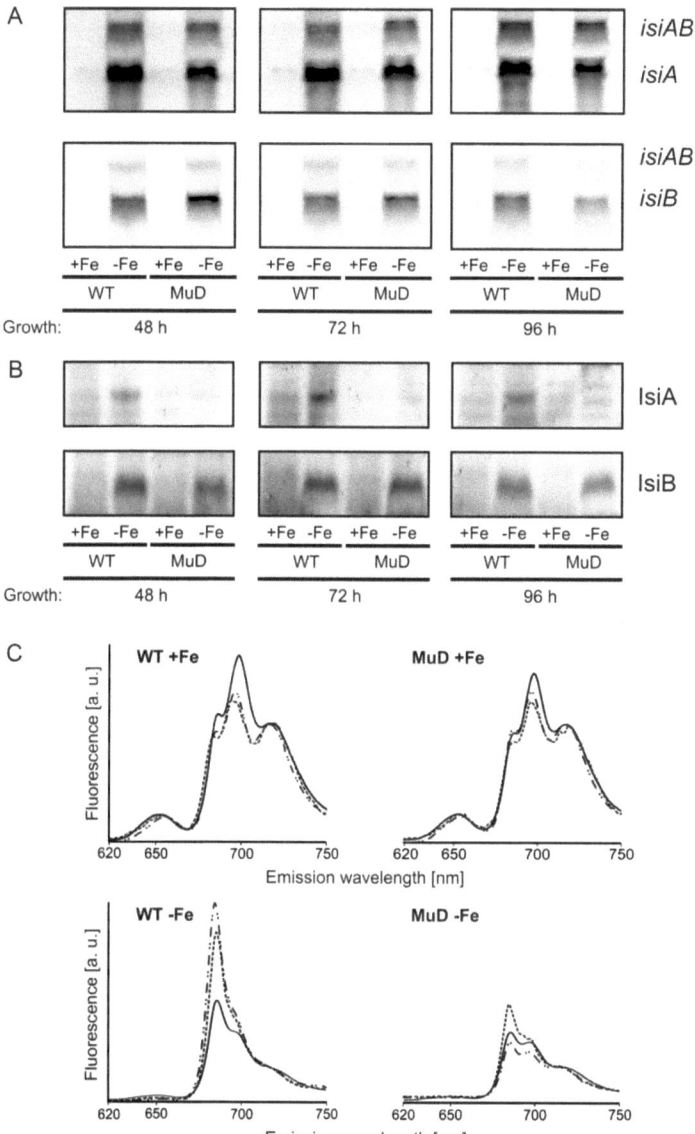

Figure 3.34: IsiA and IsiB (flavodoxin) protein expression and 77 K pigment fluorescence emission spectra of S. elongatus PCC 7942 WT and mutant MuD. Cells were grown as given in legend to Figure 3.33. **(A)** Northern blot analysis of isiA and isiB transcription with 10 µg of total RNA. Detection with gene-specific Dig-dUTP labeled probes. **(B)** Immunoblot analysis of IsiA and IsiB proteins in cell-free extracts corresponding to 100 µg proteins of cells grown for 48, 72, and 96 h in iron-sufficient or iron-deficient BG11 medium. **(C)** 77K pigment fluorescence emission spectra of intact S. elongatus PCC 7942 WT and mutant MuD cells. (—) growth for 48 h; (--) growth for 72 h; (-..-) growth for 96 h. For details see Materials and Methods. In contrast to IsiB, IsiA expression in MuD did not resemble the corresponding isiA transcription.

3.6 Comparative analysis of growth as well as IdiC, IdiB, and IdiA expression in *S. elongatus* PCC 7942 WT, the *idiC*-merodiploid mutant MuD, and the IdiB-free mutant K10

The genes *idiC* and *idiB* are located together in a single operon (see chapter 3.1). Previously, a fully-segregated IdiB-free *S. elongatus* PCC 7942 mutant, named K10, was constructed (Michel et al. 1999), while attempts to insertionally inactivate *idiC* merely led to *idiC*-merodiploid mutants. For a first comparative analysis of both mutant strains and *S. elongatus* PCC 7942 WT, growth as well as the expression of IdiB, IdiC, and the *idiB* regulated protein IdiA was investigated under iron-sufficient as well as iron-deficient growth conditions.

3.6.1 Growth of *S. elongatus* PCC 7942 WT, mutant K10 and mutant MuD under iron-sufficient and iron-depleted conditions

S. elongatus PCC 7942 WT, the IdiB-free mutant K10, and the *idiC* merodiploid mutant MuD were cultivated in regular or iron-deficient BG11 medium for 24, 48, 72, and 96 h. Cultures were inoculated with an optical density at 750 nm of 0.4. The corresponding growth curves and the phenotypical appearance were recorded (see Figure 3.35).

The growth of mutant K10 in iron-sufficient BG11 medium was comparable to that of WT during the initial growth phase and slightly decreased during prolonged growth in regular BG11 medium. In contrast, growth of mutant MuD was significantly reduced compared to that of WT and K10 under iron-sufficient conditions. The colour of the corresponding cultures showed a minor reduced pigment content compared to the appearance of WT und mutant K10 cells. The differences became more distinct under iron-depleted growth conditions. Iron-starved mutant K10 cells also showed a slightly reduced growth rate as compared to that of WT, but the cells were much more bleached, especially after prolonged growth. Growth of MuD in iron-deficient BG11 medium was barely detectable and the cultures were strongly bleached resulting in a yellowish colour. Thus, it can be assumed that mutant MuD is much more affected in its iron-dependent metabolic functions than mutant K10.

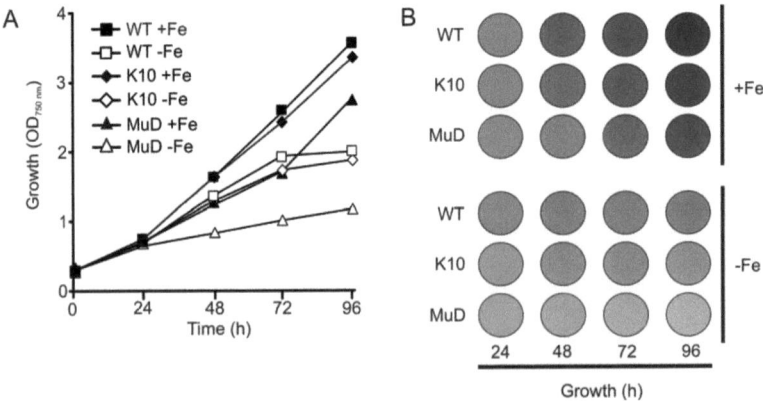

Figure 3.35: Growth (**A**) and appearance (**B**) of *S. elongatus* PCC 7942 WT cells, the IdiB-free mutant K10, and the *idiC* merodiploid mutant MuD. Cells were grown with regular or iron-depleted BG11 medium for 24, 48, 72, and 96 h. Under iron-deficient growth conditions, MuD seemed to be more severely affected than mutant K10.

3.6.2 Expression of the iron-regulated proteins IdiA, IdiB, and IdiC in S. elongatus PCC 7942 WT, mutant K10, and mutant MuD

To investigate the expression of IdiA, IdiB, and IdiC, a comparative immunoblot analysis of *S. elongatus* PCC 7942 WT, the *idiC* merodiploid mutant MuD, and the IdiB-free mutant K10 was performed. The three strains were grown for 72 h under iron-sufficient or iron-deficient conditions. Cell-free extracts corresponding to 50 µg protein were subjected to SDS PAGE and immunoblotting. Detection was performed with antisera against IdiC, IdiB, and IdiA. To maximise differences in detection of IdiC content, blots were developed with colorimetric AP staining. Immuno detection with the anti-AtpA antiserum was applied as an additional loading control, since the expression is only mildly affected by the applied iron-starved growth conditions (Exss-Sonne et al. 2000; Michel et al. 2003).

It became obvious that in all three investigated strains only minor amounts of these three investigated proteins were present under iron-sufficient growth conditions (Figure 3.36). In WT, expression of all three proteins was highly up-regulated under iron starvation. In mutant MuD low amounts of IdiC and IdiB were detected and as a consequence of the low amount of IdiB, only low amounts of IdiA were detected. In contrast, the IdiB-free mutant contained no detectable IdiB protein and thus, no IdiA was found to be present. However, a substantial amount of IdiC was detected in the IdiB-free mutant K10.

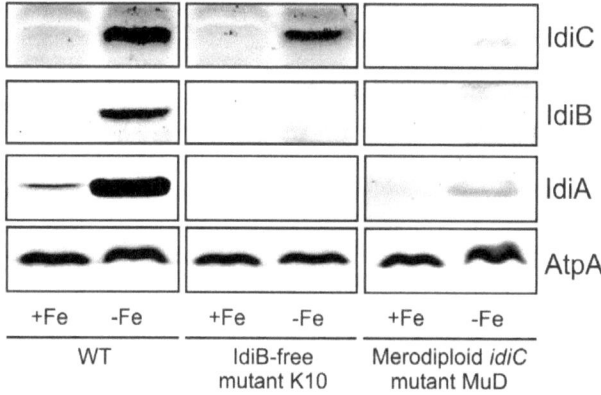

Figure 3.36: Analysis of IdiC, IdiB, and IdiA protein expression in *S. elongatus* PCC 7942 WT, the merodiploid *idiC*-mutant MuD, and the fully segregated IdiB-free *S. elongatus* PCC 7942 mutant K10 (Michel et al. 1999). The strains were grown for 72 h under iron-sufficient or iron-deficient conditions. Cell-free extracts corresponding to 50 µg protein were subjected to SDS PAGE and immunoblotting. Immunoblots were performed with the anti-IdiA (dilution 1:2,000), the anti-IdiB (dilution 1:300), the anti-IdiC (dilution 1:500), and the anti-AtpA antiserum (dilution 1:1,000). While mutant K10 showed an increased IdiC content under iron starvation, mutant MuD revealed no significant iron-dependent IdiB expression (Pietsch et al. 2007).

The results imply that despite the insertional inactivation of *idiB*, still a dicistronic message consisting of *orf6* and *idiC* was transcribed and translated in mutant K10. In addition, the results reflect the gene arrangement of the *idiB* operon. Since the gene *idiC* is located directly upstream of *idiB* and the operon is first transcribed in a primary transcript (Yousef et al. 2003), hardly any IdiB protein could be detected in MuD. The insertionally inactivated *idiC* alleles in mutant MuD are interrupted by Ω transcriptional terminators thus, preventing any efficient *idiB* transcription. In contrast, an effective *idiC* transcription was still possible, given that *idiB* was inactivated but generating reduced amounts of IdiC expression in iron-starved mutant K10 cells as compared to WT.

3.7 Comparative transcript profiling of iron-dependent regulated genes in *S. elongatus* PCC 7942 WT, the *idiC*-merodiploid mutant MuD, and the IdiB-free mutant K10

To obtain a profound view on the complex regulatory network involved in acclimation to iron limitation in *S. elongatus* PCC 7942, DNA microarray analyses with cells grown in the presence or absence of iron in BG11 medium were performed. Such microarray analyses of the genome-wide transcriptional response to iron starvation had already been performed previously for *Synechocystis* sp. PCC 6803 (Singh et al., 2003). In contrast to *Synechocystis* sp. PCC 6803 which can either grow photoautotrophically or photoheterotrophically, *S. elongatus* PCC 7942 represents an obligate photoautotrophic cyanobacterium. Since *S. elongatus* PCC 7942 is only capable of this mode of growth, it has been tempting to speculate that it might have developed more effective mechanisms to maintain its oxygenic photosynthetic lifestyle under iron starvation than the metabolically more versatile strain *Synechocystis* sp. PCC 6803. To identify iron starvation-induced gene transcription, a novel whole-genome microarray for *S. elongatus* PCC 7942 WT, which consisted of a total of 2,898 spotted 70-mer oligonucleotides, was used (Pietsch et al. 2007). Moreover, the IdiB-free *S. elongatus* PCC 7942 mutant K10 was investigated to identify novel members of an IdiB regulon and the *idiC*-merodiploid *S. elongatus* PCC 7942 mutant MuD. While it was possible to successfully insertionally inactivate the gene encoding the transcription factor IdiB (Michel et al. 1999), the *idiC*-insertionally inactivated mutant never showed full segregation (see chapter 3.5.1). Because of the suggested essential function of IdiC for the viability of *S. elongatus* PCC 7942, mutant MuD was investigated to unravel the effects of a highly reduced amount of IdiC in addition to a low content of IdiB (see chapter 3.5.4) on the overall transcriptome during acclimation to iron-depleted growth conditions.

In the DNA microarray experiments presented here, the transcriptome of *S. elongatus* PCC 7942 WT grown under iron-deficient conditions for 24 and 72 h was compared to that of WT when grown under iron-sufficient conditions for 24 and 72 h. Moreover, the transcriptomes of mutant K10 and mutant MuD, when grown for 72 h under iron-deficient conditions, were compared to those of mutant K10 and mutant MuD when grown under iron-sufficient conditions. The diagrams in Figure 3.37 give an overview on the number of differentially-regulated genes. The major changes in transcript abundance of selected genes are given in Table 3.6. Table 3.6 in combination with supplementary Table 1 (see chapter 8.1) lists the entire number of significantly transcriptionally-regulated genes. In *S. elongatus* PCC 7942 WT grown for 24 h under iron-limited condition, the steady-state transcript level of 50 genes was increased due to iron limitation, while the steady-state transcript level of 10 genes was down-regulated. After 72 h of iron depletion, 64 transcripts were found at elevated levels, while the steady-state transcript level of 24 genes diminished significantly at the same time. In mutants K10 and MuD, 42 and 60 transcripts were found in increased levels, while 21 and 60 decreased transcript levels were found after 72 h of iron-deficient growth (see Figure 3.37 and Table 3.6).

Results

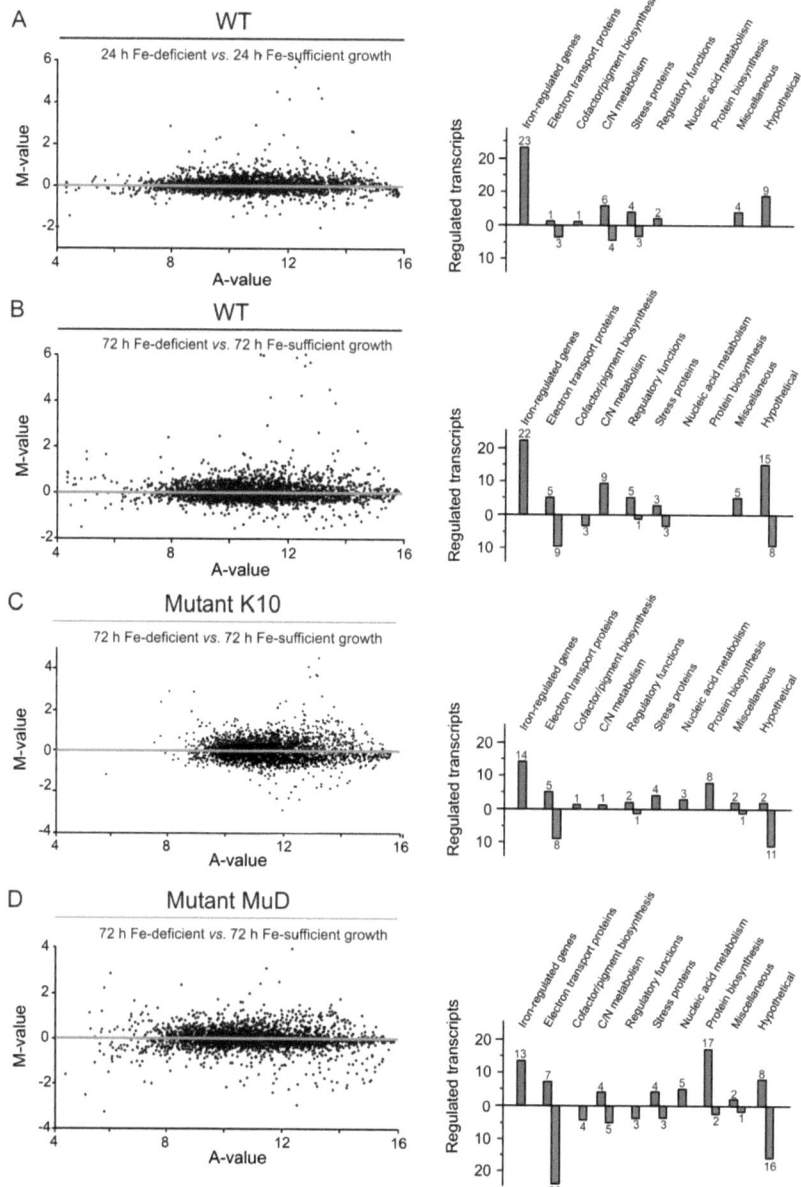

Figure 3.37: Scattered plots of differentially-regulated transcripts (left) and distribution of the genes in different metabolic categories (right) from *S. elongatus* PCC 7942 WT, the IdiB-free mutant K10, and the *idiC*-merodiploid mutant MuD grown for 24 or 72 h with BG11 medium in the presence or absence of iron. Increased transcript levels are given in red, decreased transcript levels are given in green. **(A)** *S. elongatus* PCC 7942 WT 24 h iron-deficient growth *vs. S. elongatus* PCC 7942 WT 24 h iron-sufficient growth. **(B)** *S. elongatus* PCC 7942 WT 72 h iron-deficient growth *vs. S. elongatus* PCC 7942 WT 72 h iron-sufficient growth. **(C)** IdiB-free mutant K10 72 h iron-deficient growth *vs.* IdiB-free mutant K10 72 h iron-sufficient growth. **(D)** *idiC*-merodiploid mutant MuD 72 h iron-deficient growth *vs. idiC*-merodiploid mutant MuD 72 h iron-sufficient growth.

3.7.1 Detection of transcripts of major iron-regulated clustered genes

In total, six regions on the chromosome of *S. elongatus* PCC 7942 with clusters of genes were identified, whose transcripts significantly accumulated in response to iron-deficient growth conditions. The corresponding genes are listed in Table 3.6. These gene clusters contain the genes *irpA* and *irpB*, the *fut* genes, the *suf* genes, the *isiABC* genes, the genes *idiB* and *idiC*, and the *ackA/pgam* genes. The structures of these gene regions are shown in Figure 3.38.

Table 3.6 contains the evaluated data of three biological and two technical replicates including a dye-swap experiment. The fold change value is calculated as $\log_2^{M\text{-value}}$ of M-values with corresponding p-value ≤0.051. M-values >-0.9 and <+0.9 indicate no significant change in transcriptional levels. These values correspond to a fold change of ≤1.87 and ≥0.53. Significantly increased or decreased transcript levels are printed in bold letters.

1. The highest increase in transcript abundance during iron limitation was observed for *irpA* and the transcript of gene *1461*, which we named *irpB* (see Table 3.6 A). The gene *irpB* is located immediately down-stream of *irpA* and is transcribed in the same direction. Both genes overlap by 5 bps. Thus, *irpA* and *irpB* constitute a dicistronic operon (see Figure 3.38 A). Previously, it has been suggested that IrpA is located in an operon and that the genes of this operon encode proteins of an iron acquisition system (Reddy et al. 1988). IrpA is a protein of 38.6 kDa and is located in the cytoplasmic membrane (Reddy et al. 1988). The gene *irpB* encodes a protein of 49.3 kDa and belongs to the multiheme Cyt c-type cytochrome family with two CXXC heme-binding sites (InterProScan). A role of IrpAB in iron acquisition is supported by the fact that immediately downstream of *irpA* the genes *1463* and *1464* are located that are transcribed in opposite direction to *irpAB*. These genes encode proteins with similarity to SomB(1) and SomA(1). SomA(1) and SomB(1) are outer membrane proteins that form porin-like β-barrel structures (Hansel et al. 1998), which may change the permeability and selectivity of the outer membrane as a diffusion barrier. The transcript of *somB(1)*, but not of *somA(1)*, was also found at increased concentration in the transcriptome of iron-depleted WT cells. In addition, the transcript of gene *1607*, which is located in a different region on the chromosome, and which encodes a SomA(2)-similar protein as well as the transcript of gene *2421* for a Ftr1-similar protein, were found to be substantially increased during iron limitation. Ftr1 functions as a permease of a high-affinity iron uptake system, first identified in yeast (Larrondo et al. 2007; Stearman et al. 1996). Ftr1 lies adjacent to the *ackA/pgam* operon and is transcribed in opposite direction to this operon (see Figure 3.38 F). I suggest that the proteins IrpA, IrpB, SomB(1), SomA(2), and Ftr1 represent a novel iron acquisition system in *S. elongatus* PCC 7942, whose expression is regulated by the transcriptional activator IdiB (see later in the text). IrpA- or IrpB-similar proteins are not present in all so far sequenced cyanobacterial genomes (NCBI database November 2007), while several Som-like proteins and a Ftr1-similar protein (e.g. Slr0964 in *Synechocystis* sp. PCC 6803) are present.

Table 3.6: (part **A**) List of major iron-regulated genes in *S. elongatus* PCC 7942 WT, the IdiB-free mutant K10, and the *idiC*-merodiploid mutant MuD in response to growth for 24 (WT only) or 72 h (WT, K10, and MuD) with iron-deficient *vs.* iron-sufficient BG11 medium. JGI ORFs correspond to the JGI annotation. Common gene names are given in the second column. "NA" means that the gene is not annotated in the JGI annotation system. "ND" not determined. Significantly increased or decreased transcript levels are printed in bold letters.

JGI ORF	Gene	Annotated protein function	Fold change			
			Growth -Fe *vs.* +Fe		Growth -Fe *vs.* +Fe	
			WT 24 h	WT 72 h	K10 72 h	MuD 72 h
Transcripts of major iron-regulated and clustered genes						
1462	irpA	Iron-regulated protein A	**75.58**	**62.68**	0.82	1.21
1461	irpB	Multiheme *c*-type cytochrome family protein with two heme-binding sites	**51.63**	**46.21**	1.04	0.89
1463	somB(1)	Major outer membrane protein probably forming porin-like β-barrel structure and which might also connect to the S-layer	**22.78**	**17.87**	1.38	0.95
1607	somA(2)	Major outer membrane protein, see above	**6.11**	**4.53**	0.61	0.89
2421	ftr1	Ftr1-similar protein, part of a high-affinity Fe^{2+} uptake system	**2.85**	**3.27**	1.08	1.22
1406	futC	Iron (III) transport ATP-binding protein	**2.07**	1.65	1.44	1.42
1407	futB	Iron (III) ABC transporter permease	**5.46**	**3.63**	**2.55**	**5.35**
1408	mapA	Membrane-associated protein A, partly resembles type 12 methyltransferases and periplasmic solute-binding proteins	1.37	1.44	1.50	**2.19**
1409	futA2	Iron (III) transport substrate-binding protein	**2.50**	**2.08**	**2.55**	1.50
1733	sufR	Repressor of the *suf* regulon	**2.01**	**2.27**	**2.79**	**1.91**
1734	ftrC	Ferredoxin:thioredoxin reductase catalytic subunit β-chain	**1.86**	**2.00**	**2.19**	**2.07**
1736	sufC	[Fe-S]-assembly ATPase SufC	**2.23**	**2.55**	**5.43**	**3.95**
1737	sufD	[Fe-S]-assembly protein SufD	**2.00**	**2.30**	**4.41**	**3.05**
1738	sufS	Cysteine desulfatase, NifS-similar, involved in formation of [Fe-S] centres	**2.30**	**2.57**	**4.99**	**3.02**
1739	merR	MerR-like protein containing a HTH DNA-binding motif	**1.92**	**2.04**	**4.93**	**2.33**
NA	0017	Hypothetical 4.2 kDa protein with signal peptide	**7.52**	**7.62**	**3.39**	**1.95**
1542	isiA	Iron stress-induced protein A or CP43', formation of membrane-integral light harvesting antenna around trimeric photosystem I	**25.99**	**22.79**	**23.59**	**11.00**
1541	isiB	Flavodoxin, soluble electron transport protein, in part replaces ferredoxin under iron starvation	**61.82**	**51.63**	**13.45**	**5.66**
1540	isiC	Putative hydrolase with typical αβ-fold of hydrolases	**68.60**	**63.35**	**17.39**	**8.64**
2175	idiA	Iron deficiency-induced protein A, modifies and protects photosystem II against selected stresses	**6.24**	**6.50**	0.68	1.00
2174	idiB	Iron deficiency-induced protein B, positively acting transcription factor of IdiA and the IdiB regulon	**17.88**	**18.89**	**6.41**	1.18
2173	idiC	Iron deficiency-induced protein C, suggested to participate in photosynthetic cyclic electron transport	**18.50**	**15.24**	**12.91**	**16.34**
2172	orf6	Gene immediately upstream of *idiC*, encodes a protein of unknown function	ND	1.26	**12.55**	**2.27**

Table 3.6: Continued (part **B**).

JGI ORF	Gene	Annotated protein function	Fold change			
			Growth -Fe vs. +Fe		Growth -Fe vs. +Fe	
			WT 24 h	WT 72 h	K10 72 h	MuD 72 h
Transcripts of major iron-regulated and clustered genes						
2079	ackA	Acetate kinase, production of acetate from acetyl-phosphate with synthesis of ATP	**5.03**	**3.14**	1.06	0.97
2078	pgam	Phosphoglycerate mutase, transfers phosphate groups within glycerate and converts 3- to 2-PGA	**3.05**	**2.43**	1.38	1.06
Transcripts of genes encoding electron transport-related proteins (photosynthesis and respiration)						
0679	psbB	CP47 light-harvesting antenna protein of photosystem II	0.91	0.87	0.71	**0.54**
0656	psbC	CP43 light-harvesting antenna protein of photosystem II	0.88	1.15	0.84	**0.46**
0294	psbO	Manganese- and calcium-stabilising protein of photosystem II	0.63	**0.52**	**0.42**	**0.42**
0696	psbT	Small photosystem II protein; involved in stabilisation of photosystem II dimers and recovery from photodamage	0.94	1.02	0.84	**0.46**
2049	psaA	Photosystem I reaction centre core protein A	0.80	0.92	0.62	**0.43**
2048	psaB	Photosystem I reaction centre core protein B	0.75	0.79	0.61	**0.38**
0535	psaC	Photosystem I reaction centre subunit C, F_A and F_B [4Fe-4S]-containing protein on the stromal surface of photosystem I	0.93	0.91	0.75	**0.35**
0407	psaK(1)	Photosystem I reaction centre subunit X, hydrophobic subunit of unknown function	0.91	**0.49**	**0.30**	**0.34**
0920	psaK(2)	Photosystem I reaction centre subunit X, hydrophobic subunit of unknown function	1.04	**0.53**	**0.31**	**0.34**
2342	psaL	Photosystem I reaction centre subunit L, trimerisation of photosystem I monomers	0.97	**0.56**	**0.34**	**0.30**
1249	psaJ	Photosystem I reaction centre subunit IX, hydrophobic subunit close to PsaF	0.90	0.86	0.57	**0.29**
2343	psaI	Photosystem I reaction centre subunit VIII, crucial role in aiding normal structural organisation of PsaL	0.99	**0.50**	**0.35**	**0.27**
1250	psaF	Photosystem I reaction centre subunit III, small subunit of unknown function	1.00	0.65	**0.37**	**0.26**
1231	petA	Apocytochrome f	**0.47**	**0.50**	0.95	0.86
2331	petB	Cytochrome b_6	0.86	1.24	0.79	0.68
1232	petC	Rieske protein	**0.48**	**0.49**	0.98	0.82
1630	petJ	Apocytochrome c_6 precursor (Cyt c_{553})	0.57	**0.48**	0.56	0.61
2332	petD	Cytochrome b_6/f complex subunit IV	0.63	0.76	0.71	**0.50**
1439	ndhD	NADH dehydrogenase subunit IV (NdhD2)	**2.11**	**6.15**	**3.48**	**2.79**
1767	cydA	Cytochrome bd oxidase subunit I	1.27	**2.08**	**2.04**	**1.89**
2601	ctaA	Cytochrome oxidase assembly protein, required for assembly of Cyt aa_3 oxidase	1.55	1.36	1.61	**2.04**
2602	ctaC	Cytochrome aa_3 oxidase subunit II	1.52	1.55	**2.64**	**2.17**
2604	ctaE	Cytochrome aa_3 oxidase subunit III	1.82	**2.20**	**4.29**	**2.20**
0201	ccoO	Cytochrome oxidase cytochrome c subunit	1.26	1.73	**1.97**	1.44

Table 3.6: Continued (part **C**).

JGI ORF	Gene	Annotated protein function	Fold change			
			Growth -Fe vs. +Fe		Growth -Fe vs. +Fe	
			WT 24 h	WT 72 h	K10 72 h	MuD 72 h
Transcripts of genes encoding electron transport-related proteins (photosynthesis and respiration)						
0202	ccoN	Cytochrome oxidase, cb-type cytochrome oxidase subunit I	1.24	**1.92**	1.82	1.39
0814	2124	Putative 7x Fe ferredoxin with [3Fe-4S] and [4Fe-4S] cofactors	**0.51**	**0.48**	0.87	0.94
1649	0140	Ruberythrin and rubredoxin-type [4Fe-4S]-like protein, putatively involved in electron transfer reactions, sometimes replacing ferredoxins in electron transport.	0.83	0.93	1.80	**2.04**
0327	apcA	Allophycocyanin α-subunit	0.77	0.87	0.63	**0.26**
2158	apcB(1)	Allophycocyanin β-subunit (1)	1.05	0.82	0.48	**0.48**
0328	apcE	Phycobilisome anchor protein	0.66	**0.50**	0.43	**0.21**
0325	apcI	Allophycocyanin linker protein	0.89	**0.42**	0.29	**0.23**
1053	cpcA	Phycocyanin α-subunit (1)	0.77	0.99	0.95	**0.46**
1048	cpcA	Phycocyanin α-subunit (2)	1.31	**3.14**	0.92	**4.63**
1052	cpcB(1)	Phycocyanin β-subunit (1)	0.76	1.03	0.82	**0.28**
1047	cpcB(2)	Phycocyanin β-subunit (2)	0.76	1.05	0.80	**0.29**
1050	cpcI(1)	33 kDa phycocyanin linker protein	0.83	0.81	0.72	**0.23**
1051	cpcI(2)	33 kDa phycocyanin linker protein	0.81	0.63	0.61	**0.36**
1049	cpcH	Rod-rod linker protein	0.75	0.78	0.64	**0.25**
2030	cpcG	Phycobilisome rod-core linker polypeptide	0.93	0.93	0.72	**0.48**
Transcripts of genes encoding proteins of carbon- and nitrogen metabolism						
2079	ackA	Acetate kinase, production of acetate from acetyl-phosphate with synthesis of ATP	**5.03**	**3.14**	1.06	0.97
0650	nat	N-Acetyltransferase	**4.74**	**4.72**	1.36	0.50
2078	pgam	Phosphoglycerate mutase, transfers phosphate groups within glycerate molecules, converts 3-PGA to 2-PGA	**3.05**	**2.43**	1.38	1.06
1608	0094	Mannose-1-phosphate guanylyltransferase/mannose-6-phosphate isomerase	**2.04**	**2.17**	1.30	0.85
1609	nrdJ	Ribonucleoside triphosphate reductase, adenosyl-cobalamine-dependent enzyme	1.78	**2.36**	1.79	1.06
1072	cobO	Cobalamine adenosyl transferase, involved in adenosyl cobalamine biosynthesis	1.46	**2.06**	1.22	1.26
1585	0069	N-acetylmuramoyl L-alanine amidase	1.39	**2.00**	1.22	1.35
2388	oxdC	Oxalate decarboxylase with cupin-like β-barrels	1.11	0.79	0.58	**0.47**
1240	nirA	Ferredoxin-nitrite reductase	0.79	0.82	0.59	**0.45**
1239	nrtA	ABC-type nitrate transporter subunit A	0.57	0.88	0.60	**0.44**
1237	nrtC	ABC-type nitrate transporter subunit C	0.63	1.02	0.76	**0.54**
2529	gifB	Hypothetical 11.6 kDa protein, similar to Sll1515 from *Synechocystis* sp. PCC 6803 to glutamine synthetase inactivating factor IF17	**3.29**	**2.27**	1.42	**2.36**
2150	0687	Linear amide C-N hydrolase, choloylglycine hydrolase, member of the NTN hydrolase family	**2.17**	**1.95**	1.23	1.19

Table 3.6: Continued (part D).

JGI ORF	Gene	Annotated protein function	Fold change			
			Growth -Fe vs. +Fe		Growth -Fe vs. +Fe	
			WT 24 h	WT 72 h	K10 72 h	MuD 72 h
Transcripts of genes encoding proteins of carbon- and nitrogen metabolism						
1513	dxr	1-deoxy-D-xylulose 5-phosphate reductoisomerase, involved in terpenoid orisoprenoid biosynthesis	0.75	0.74	0.85	**0.48**
1562	draG	ADP-ribosylglycohydrolase dinitrogenase reductase activating glycohydrolase	0.95	1.80	**2.99**	**3.29**
1713	mocD	Hydrocarbon oxygenase-similar protein	0.81	1.21	1.14	**2.00**
1438	pmgA	Photomixotrophic growth-related protein A homolog	1.60	**1.99**	1.32	1.39
2043	speH	S-adenosyl methionine decarboxylase, involved in spermidine biosynthesis	0.86	1.45	1.84	**2.77**
1513	dxr	1-deoxy-D-xylulose 5-phosphate reductoisomerase, involved in terpenoid orisoprenoid biosynthesis	0.75	0.74	0.85	**0.48**
0510	serB	Haloacid dehydrogenase-like hydrolase	0.84	1.02	1.59	**2.17**
2107	cynA	Periplasmic-binding protein	**0.39**	0.84	1.43	1.16
2106	cynB	Sulfonate transport system permease protein	**0.40**	0.89	1.41	1.55
2105	cynD	ATP-binding protein of sulfonate transport system	**0.49**	0.93	1.49	1.29
2104	cynS	Cyanase, detoxification of cyanate (N≡C-O⁻)	**0.24**	0.78	1.20	1.09
Transcripts of genes encoding general stress proteins						
1813	htpG	Heat shock protein HSP90	**2.33**	**2.12**	1.56	0.90
2313	groL	GroL chaperonine HSP60	**2.03**	1.23	**2.04**	1.01
2314	groS	GroS chaperonine HSP10	**2.27**	1.27	**2.04**	1.13
2306	dnaJ-like	Protein similar to the C-terminus of DnaJ (HSP40) lacking three conserved domains of DnaJ proteins	0.88	0.89	0.74	**0.45**
2401	hspA	Molecular chaperone of the HSP20 family	1.13	1.77	**2.00**	**2.02**
0801	sodB	Fe-superoxide dismutase (Fe-SOD)	**4.00**	**0.37**	0.53	**0.34**
1656	katG	Catalase peroxidase	0.84	**0.43**	0.51	**0.30**
2309	aphC	Alkyl hydroperoxide reductase C, 2-Cys peroxiredoxin-type protein	1.24	1.78	1.88	**2.03**
1290	2659	Metallothionine-similar protein	**0.21**	**0.34**	0.66	1.11
0243	hliC	High light-induced protein C, LHC-like protein Lhl4	1.25	**3.61**	**4.53**	**2.83**
2127	nblA	Non-bleaching protein A, phycobilisome-degradation trigger protein	1.58	**2.66**	1.83	**2.07**
1635	somB(2)	Major outer membrane protein probably forming porin-like-barrel structure and putative connection to S-layer	**0.50**	0.57	1.32	1.25

2. Transcripts for a second iron acquisition system, the Fut system (Katoh et al. 2001a; Katoh et al. 2001b), were also found to be transcribed at elevated levels in *S. elongatus* PCC 7942 (*futA2*, *futB*, and *futC*) (see Table 3.6 A and Figure 3.38). Among these transcripts, the highest increase was seen for the steady-state level of *futB* mRNA, encoding a putative iron-(III)-transporter permease. Since the gene *mapA*, encoding a 34 kDa protein (Webb et al. 1994), is located between *futB* and *futA2*, MapA may also play a role in iron acquisition. Although *futB*, *mapA*, and *futA2* are arranged in line on the chromosome and transcribed in the same direction, the increase in their steady-state mRNA levels were found to be substantially different under iron-limiting growth conditions suggesting that their transcript stability is rather different. The gene *futC* is located upstream of *futB*, but is transcribed in opposite direction (see Figure 3.38 B).

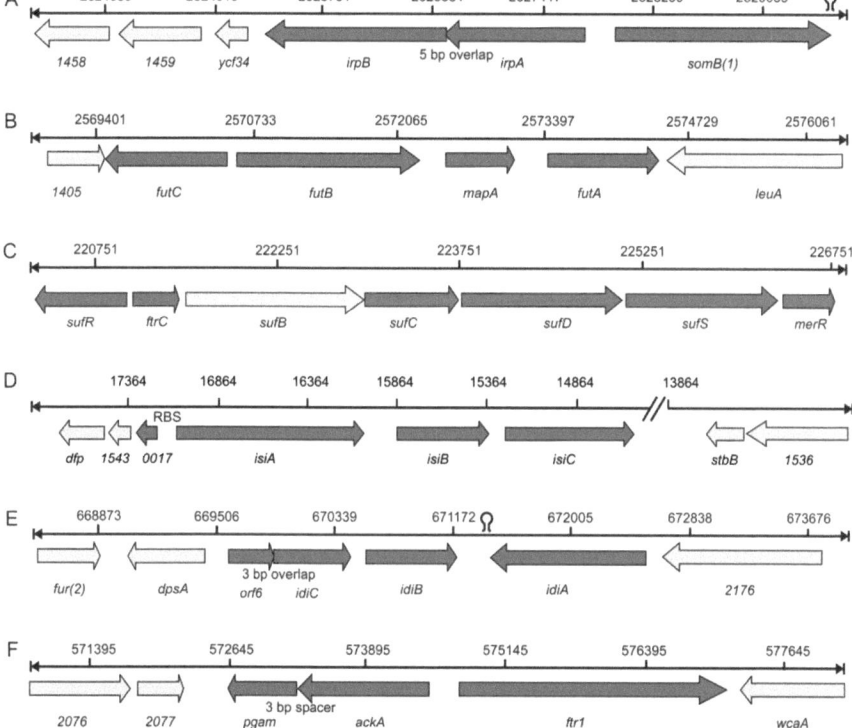

Figure 3.38: Partial map of the *S. elongatus* PCC 7942 chromosome with genes arranged in sequence that transcriptionally respond to the iron status of the cell: **(A)** *irpAB* region, **(B)** *fut* region, **(C)** *suf* region, **(D)** *isiAB* region, **(E)** *idiCB* region, and **(E)** the *ackA* region. Transcription of *idiA* has previously been shown to be regulated by IdiB (Michel et al. 2001). The results suggest that the *irpAB* regulon and the acetate kinase regulon are also regulated by IdiB. The *isiAB* operon is regulated by the transcriptional repressor Fur (Ghassemian and Straus 1996). The *suf* operon is assumed to be regulated by the repressor SufR (Wang et al. 2004). The transcriptional regulator(s) for the *idiBC* and the *fut* operon are still unknown. The genes given in red were up-regulated in *S. elongatus* PCC 7942 WT in the course of iron starvation.

As for futB, the steady-state transcript pool for the futC mRNA was up-regulated under conditions of iron starvation. The cyanobacterial Fut system is closely related to the well-characterised bacterial Sfu-, Hit-, and Fbp iron uptake systems (Angerer et al. 1990; Chen et al. 1993; Nowalk et al. 1994; Sanders et al. 1994). Since the increase in transcript level for the four fut genes was different, it remained unclear, whether three of these genes indeed form an operon like in other eubacteria or represent a regulon. A transcriptional regulator for the Fut system of S. elongatus PCC 7942 has so far not been identified. In Synechocystis sp. PCC 6803, a function in iron acquisition has been proven for FutA1 (Slr1295), FutA2 (Slr0513), FutB (Slr0327), and FutC (Sll1878) (Katoh et al. 2001a; Katoh et al. 2001b). These proteins represent an ATP-binding cassette (ABC)-type ferric iron transporter. FutA1 (Koropatkin et al. 2007) and FutA2 are iron-binding proteins, and FutB and FutC contain nucleotide-binding motifs and belong to the ABC-transporter family of inner-membrane-bound and membrane-associated proteins, respectively (Katoh et al. 2001a; Katoh et al. 2001b). In Synechocystis sp. PCC 6803, FutA2 is predominantly located in the periplasm (Fulda et al. 2000; Fulda et al. 1999), while FutA1 is mainly detected in the thylakoid membrane fraction co-purifying with PS II (Tölle et al. 2002). The localisation of FutA2 in S. elongatus PCC 7942 has not yet been investigated. The protein MapA has been shown to be predominantly located in the cytoplasmic membrane, but has also been detected in the thylakoid membrane of iron-depleted S. elongatus PCC 7942 cells (Webb et al. 1994). The N-terminal part of MapA has similarity to the chloroplast envelope protein E37, while the C-terminal part resembles bacterial iron acquisition proteins. Like IrpA and IrpB (see above), MapA has no counterpart in any of the so far sequenced and annotated cyanobacterial genomes (NCBI database, October 2007).

3. The Suf [Fe-S] assembly system of S. elongatus PCC 7942 is assumed to function as an auxiliary [Fe-S] assembly system besides the house-keeping Isc system and most likely facilitates the assembly and/or repair of the oxygen-labile [Fe-S] clusters under conditions of oxidative stress and iron limitation (Balasubramanian et al. 2006; Nachin et al. 2003; Wang et al. 2004). Selected transcripts of this system were up-regulated in S. elongatus PCC 7942 during iron limitation. In detail, an increase in the steady-state transcript level was seen for the transcripts of sufC, sufD, and sufS but not for sufB (see Table 3.6 A). An increase was also observed for the transcript of sufR (1733) as well as for ftrC, and for a gene encoding a MerR-like HTH-type transcription factor. In total, this gene cluster comprises six genes, which are arranged in line and which are transcribed in the same direction (genes ftrC, sufB, sufC, sufD, sufS, and merR), and one gene (sufR), which lies adjacent but which is transcribed in the opposite direction as shown in Figure 3.38 C. The transcripts/proteins of this system have been shown to be expressed at elevated levels in several cyanobacteria during iron-limiting growth conditions. The structure of the suf operon found in S. elongatus PC 7942 is similar to those of several other cyanobacterial strains, such as Synechocystis sp. PCC 6803, Synechococcus sp. PCC 7002, and Anabaena sp. PCC 7120 (Wang et al. 2004). Since the ftrC gene is located between sufR and the sufBCDS operon in S. elongatus PCC 7942, the suf region resembles mostly that of Synechococcus sp. strain WH8102. Gene 1734 encodes a protein with similarity to the catalytic subunit of a [4Fe-4S] ferredoxin:thioredoxin reductase (FtrC), that may function in

thioredoxin-mediated redox regulation of protein function and signalling via a thiol redox control (Wang et al. 2004). SufR has been shown to be a [4Fe-4S] protein, which acts as the transcriptional repressor of the *suf* operon in *Synechococcus* sp. PCC 7002. The *suf* operon in *S. elongatus* PCC 7942 contains an additional gene for a MerR-type transcriptional regulator protein. MerR belongs to the family of HTH transcription factors including SoxR of *E. coli*, which e.g. activates the transcription of flavodoxin (Brown et al. 2003). The MerR protein of *S. elongatus* PCC 7942 is somewhat atypical, since it lacks two of four invariant cysteine residues, which have been shown to bind the [4Fe-4S] cofactor of the repressor. Whether MerR is involved in the regulation of the *suf* genes in addition to SufR in *S. elongatus* PCC 7942, is still unknown.

4. The fourth operon in *S. elongatus* PCC 7942, whose steady-state transcript level was found to be highly increased as a consequence of iron depletion, contains the genes *isiA*, *isiB*, and *isiC* (see Figure 3.38 D). The gene *isiA* encodes CP43', and *isiB* encodes flavodoxin. The function of these two proteins has been discussed in the introduction. Moreover, this operon also contains a gene that we named *isiC* and that encodes a hydrolase-like protein. This putative hydrolase with a deduced molecular mass of 26.9 kDa is 54.3% similar to the esterase Fes (18.2% identical amino acid residues, 22.1% strongly similar amino acid residues, 14% weakly similar amino acid residues) from *E. coli*. FesA hydrolyses ester bonds of internalised ferri-enterobactin siderophores (Andrews et al. 2003). The increased transcript abundance of *isiB* and *isiC* was substantially higher than that of *isiA*. Differences in the *isiA* and *isiB* steady-state transcript levels under iron-limiting growth conditions have already been described for *S. elongatus* PCC 7942 (Bagchi et al. 2007; Bagchi et al. 2003; Pietsch et al. 2007). In addition, it has also been observed that a monocistronic *isiA* message was more abundant than a dicistronic *isiAB* message under iron limitation (Straus 1994). The expression of this operon is under the control of the transcriptional repressor Fur in *S. elongatus* PCC 7942 (Ghassemian and Straus 1996). Gene *0017* is located upstream of *isiA* and encodes a putative 4.2 kDa polypeptide of unknown function. This gene is transcribed in the opposite direction, and its transcript abundance was also strongly increased under iron limitation. Thus, this particular genome region contains four genes in total, whose transcript levels increased during iron limitation. For *Synechocystis* sp. PCC 6803, it has previously been reported that the *isiAB* region contains three more genes in addition to *isiA* and *isiB* that were transcribed at elevated levels during iron-limited growth conditions (Singh et al. 2003).

5. A fifth group of transcripts, which accumulated under iron starvation, was that of *idiB* and *idiC*. The function of the corresponding proteins is explained in the introduction (Michel and Pistorius 2004; Pietsch et al. 2007; Yousef et al. 2003). The *idiBC* genes are localised in a single operon (Figure 3.38 E). Whether the third gene named *orf6* encodes a protein, remains to be investigated (Yousef et al. 2003). The gene *idiA* separates from the neighboring *idiCB* operon by a strong terminator sequence, and it is transcribed in the opposite direction. The increase in the steady-state pool of *idiB* and *idiC* mRNA was almost as high as in the case of *isiA*, while the increase of the *idiA* mRNA pool was about one third of the *idiB* level. IdiB is a helix-turn-helix-type transcription factor and regulates transcription of *idiA* (Michel et al. 2001). The iron-responsive transcriptional regulator of

the *idiCB* operon is still unknown. *Synechocystis* sp. PCC 6803 also contains an IdiA-similar protein Slr1295 (Tölle et al. 2002) named FutA1 (Katoh et al. 2001a; Katoh et al. 2001b), but lacks an IdiB-similar as well as an IdiC-similar protein. The gene *dpsA* is located upstream of the *idiCB* operon. The corresponding transcript was not found at increased levels under iron-depleted growth conditions (Michel et al. 2003). DpsA is a DNA-binding heme protein and confers resistance to oxidative stress to genomic DNA (Dwivedi et al. 1997).

6. The sixth region with iron-regulated genes arranged in sequence comprises two genes that are separated by only three base pairs and that are transcribed in the same direction. The genes encode an acetate kinase (AckA) and a phosphoglycerate mutase (Pgam). An increase of the steady-state mRNA pool of these two genes was observed during iron starvation in WT but not in mutant K10 and mutant MuD (see Table 3.6 B). The observed increase of the phosphoglycerate mutase transcript concentration implies that during iron limitation 3-phosphoglycerate (3-PGA) is in part withdrawn from the Calvin cycle to increase the rate of glycolysis, and the elevated transcript level for acetate kinase further supports the assumption that pyruvate in part becomes metabolized to acetate. This fact would imply that glycogen fermentation and increased utilisation of Calvin cycle intermediates in catabolism results in an enhanced production of acetate (Moezelaar et al. 1996; Moezelaar and Stal 1994; Steunou et al. 2006; van der Oost et al. 1989) and an additional synthesis of ATP (see Figure 3.39).

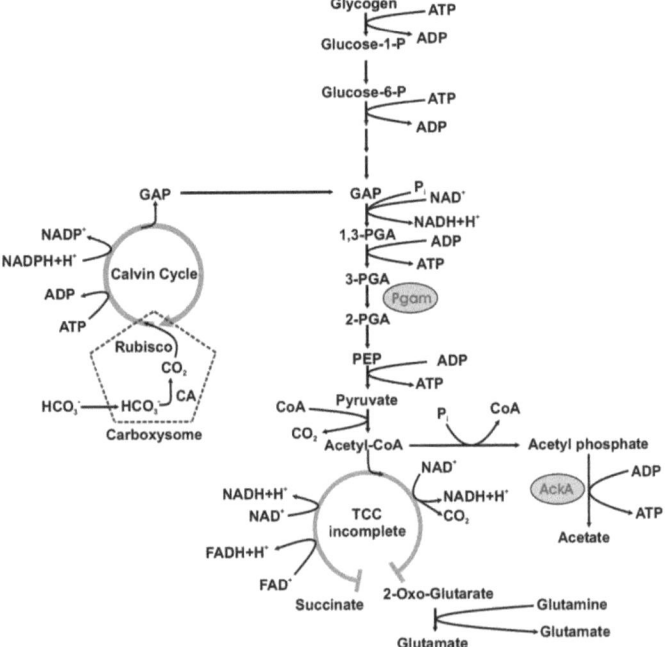

Figure 3.39: The role of Pgam and AckA in glycolysis and subtsrate-level phosphorylation via dephosphorylation of acetyl phosphate in *S. elongatus* PCC 7942.

For verification of the DNA microarray results related to the increased transcript level of the genes under iron starvation being located in an operon or regulon, Northern blots were performed with gene-specific Dig dUTP-labelled DNA probes and total RNA was isolated from S. elongatus PCC 7942 grown either with iron-sufficient or iron-deficient BG11 medium for 24 or 72 h (see Figure 3.40). An increased mRNA level under iron limitation was verified for all genes that are listed in Table 3.6 A. The highest increase was detected for the genes irpA, isiB, and isiC, which is in good agreement with the DNA microarray results. A somewhat lower increase was observed for the transcripts of the suf operon, which again agrees quite well with the obtained DN microarray data.

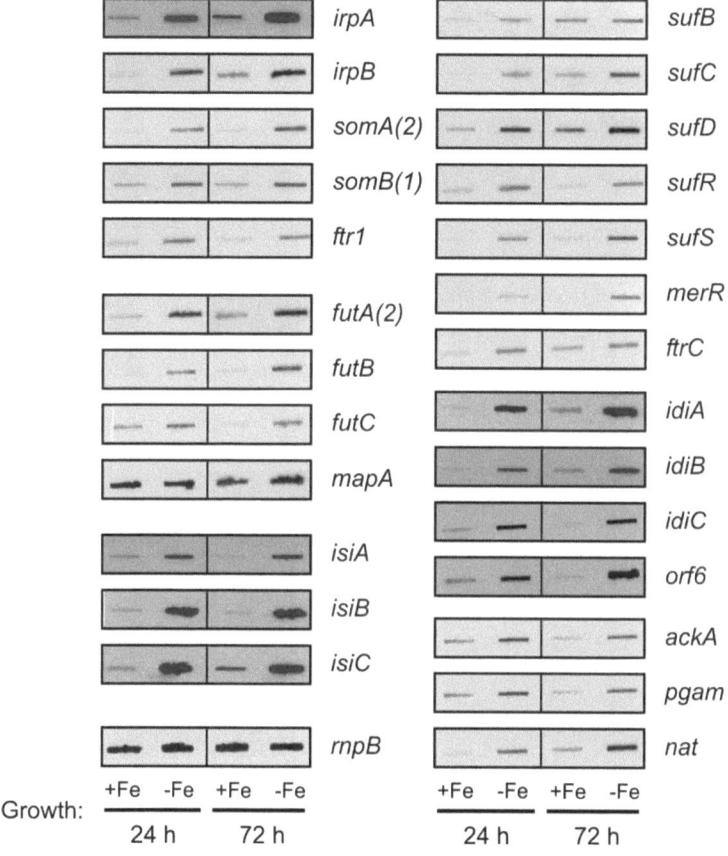

Figure 3.40: Transcript analysis of selected iron-regulated genes from S. elongatus PCC 7942 WT. Total RNA was isolated from S. elongatus PCC 7942 WT cultures grown either in the presence or absence of iron for 24 and 72 h. 2 µg RNA was vacuum-blotted on Hybond⁺™ nylon membranes. Steady-state transcript pools were detected with gene-specific Dig-dUTP-labelled DNA probes. An rnpB-specific probe was used to assure equal loading (Pietsch et al. 2007).

3.7.2 Detection of transcripts encoding electron transport-related proteins (photosynthesis and respiration)

The steady-state *psbO* transcript level for the Mn- and Ca^{2+}-stabilising protein of PS II (De Las Rivas et al. 2004), which has been assigned a regulatory function for the photosynthetic oxygen evolving activity at least in some cyanobacteria (Sherman et al. 1998; Tucker et al. 2001), showed the strongest decrease among the PS II-related transcripts (Ke 2001). In addition, the transcript concentrations for proximal antenna proteins CP47 and CP43, and also the transcript pool of PsbT, which probably contributes to stabilisation of PS II dimers, were slightly decreased (see Table 3.6 B).

Among the PS I-encoding genes (Ke 2001), the steady-state transcript level for the reaction centre proteins PsaA, PsaB, and PsaC as well as the transcripts for the auxiliary PS I proteins PsaK, PsaL, PsaJ, Psal, and PsaF were diminished under iron depletion. PsaL participates in trimerisation of PS I monomers (Chitnis and Chitnis 1993) and Psal aids the structural organisation of PsaL. This result indicates that prolonged iron limitation results in a reduction of PS I reaction centre proteins and in a decreased percentage of trimerised PS I relative to the total pool of PS I. The latter finding supports previous results showing that monomeric PS I is favoured over trimeric PS I in *S. elongatus* PCC 7942 during iron-deficient growth conditions (Ivanov et al. 2006). A similar result was also obtained from *Synechocystis* sp. PCC 6803 grown under prolonged iron limitation (Yeremenko et al. 2004).

Transcript levels of the four major proteins of the Cyt b_6/f complex, *petA*, *petB*, *petC*, and *petD* as well as the transcript for the mobile electron carrier Cyt c_{553} (PetJ) were also reduced during iron depletion. Furthermore, the transcript level for gene *0814*, encoding a protein with a putative [3Fe-4S] and a [4Fe-4S] cofactor, was detected at decreased concentrations.

In the course of iron depletion, the amount of a number of transcripts encoding subunits of the phycobilisome antenna and transcripts for enzymes involved in pigment biosynthesis substantially decreased in the transcriptome of *S. elongatus* PCC 7942 WT.

A change in the transcript level was also detected for a specific subunit of the cyanobacterial NDH-1 complex (Badger et al. 2002; Badger and Price 2003; Kaplan and Reinhold 1999; Ohkawa et al. 2002; Ohkawa et al. 2001) due to iron depletion. The amount of the *ndhD2* transcript, encoding a protein, which has a function in the NDH-1-type A complex-mediated respiration, was found at an elevated level. This finding is in agreement with the observation that iron limitation results in an enhanced respiratory and photosynthetic cyclic electron transport (Michel and Pistorius 2004) and a reduced linear electron transport activity (Ivanov et al. 2000). An increased mRNA level for subunits of various terminal oxidases of the respiratory electron transport chain (Schmetterer 1994; Vermaas 2001) was also detected. For example the steady-state mRNA level of *ctaA*, *ctaC*, and *ctaE*, encoding three subunits of the Cyt oxidase aa_3 (similar to the mitochondrial complex IV), was measured at elevated levels. In addition, the transcript levels for *cydA* (cyanide-sensitive alternative Cyt *bd*-type quinol oxidase *bd* subunit), *ccoO*, and *ccoN* (subunits of Cyt *cb*-type oxidase) were detected at increased levels. Concomitantly with an increased transcript level, an increase in the capacity of the terminal oxidase(s) would lead to an enhanced electron transfer capacity to molecular oxygen under conditions where PS II and PS I activities decline.

3.7.3 Detection of transcripts of genes encoding carbon metabolism-related proteins

A few transcript levels for carbon metabolism-related proteins were significantly increased in the transcriptome of iron-depleted *S. elongatus* PCC 7942 WT cells. The transcript level for the phosphoglycerate mutase (Pgam), an enzyme of glycolysis, increased due to iron starvation, whereas no substantial changes were measured for transcripts of other glycolytic enzymes. In addition, a significant increase of the acetate kinase transcript (*ackA*) concentration, a N-acetyltransferase transcript (*nat*), and a mannose-1-phosphate guanylyltransferase/mannose-6-phosphate isomerase transcript (*nrdJ*) were measured (see Table 3.6 C and Figure 3.40). The genes encoding Pgam and AckA are located next to each other and are transcribed in the same direction (see Figure 3.38 F).

3.7.4 Detection of transcripts of genes encoding nitrogen metabolism-related proteins

Transcript levels of mRNAs for proteins involved in nitrate/nitrite assimilation were slightly reduced in the transcriptome of iron-starved WT cells (see Table 3.6 C and D). E.g. the *nirA* transcript that encodes the ferredoxin:nitrite reductase (NIR) as well as transcripts for proteins of the nitrate/nitrite uptake system were detected at lower amounts. An increase of the steady-state mRNA level for a putative glutamine synthetase-inactivating factor similar to Sll1515 of *Synechocystis* sp. PCC 6803 was detected, suggesting that N-assimilation was reduced most likely due to the lower photosynthetic activity under iron limitation.

3.7.5 Detection of transcripts of genes encoding general stress proteins

The transcripts for several chaperones and/or heat shock proteins such as GroS, GroL, and HtpG were detected at significantly increased concentrations under iron-limiting growth conditions (see Table 3.6 D). Especially, the steady-state *hspA* transcript pool was increased after 72 h of iron-limited growth. In contrast, the mRNA level for a DnaJ-like protein of the HSP40 family was substantially decreased. The transcript concentration for the iron superoxide dismutase (Fe-SOD) (Herbert et al. 1992; Samson et al. 1994) was found to be lower in iron-depleted cells than in iron-sufficient cells. Altogether, these findings reveal that iron-independent detoxification systems compensate in part the iron-dependent parts of the cellular detoxification system. It could also suggest that the rate of superoxide anion formation was lower in PS II and PS I of iron-starved cells when the protective proteins IdiA and IsiA were expressed at highly elevated concentrations. Moreover, the transcript pool for the heme-containing catalase peroxidase KatG (Tichy and Vermaas 1999) was reduced under iron limitation, while the transcript abundance for *aphC* (gene *2309*), encoding a 2-Cys peroxiredoxin (Dietz et al. 2002; Stork et al. 2005; Tichy and Vermaas 1999), was increased. This increase only occurred under prolonged iron limitation. Under these conditions KatG may in part be replaced by a peroxiredoxin that does not require a catalytic iron cofactor. Amongst six so far identified peroxiredoxins of *S. elongatus* PCC 7942 (Stork et al. 2005), the 2-Cys peroxiredoxin is the one with the highest hydrogen peroxide-decomposing activity (Stork T., unpublished results). Such a compensatory role of catalase and a peroxiredoxin has previously been suggested e.g. for *Synechocystis* sp. PCC 6803 and *Staphylococcus aureus* (Cosgrove et al. 2007; Tichy and Vermaas 1999). Probably as a consequence of the reduced iron concentration in the cell, a metallothionine-related transcript was also down-regulated in *S. elongatus* PCC 7942, especially in the early phase of iron limitation. As expected, levels of the high light-induced protein C transcript (Huang et al. 2002), and of the *nblA* transcript, encoding a protein that is involved in phycobilisome degradation (Collier and Grossman 1994; van Waasbergen et al. 2002), were found to be significantly increased in iron-depleted cells.

3.7.6 Detection of transcripts of genes encoding regulatory proteins

Transcripts of genes *2466* and *1316*, encoding a CheY-like response regulator similar to Rre37 of *Synechocystis* sp. PCC 6803 and Ycf27 of *Guillardia theta*, and a transcription factor similar to Tlr1758 of *Thermosynechococcus elongatus* BP-1 were found at increased concentrations in the transcriptome of iron-starved *S. elongatus* PCC 7942 WT cells. Furthermore, the transcript levels for three alternative σ factors, *rpoD4* (group II σ factor), *rpoD3* (group II σ factor), and *sigF2* (group III σ factor), were changed as a result of prolonged iron depletion. Moreover, the mRNA level for an anti-σ factor antagonist-similar protein was increased, while the transcripts for the light-repressed transcript A protein (LrtA), which is suggested to either modulate cellular transcription- and/or translation activity (Samartzidou and Widger 1998) in response to illumination, was found to be decreased in the transcriptome of iron-depleted cultures.

In summary, the adaptational response of *S. elongatus* PCC 7942 WT to iron starvation comprises the transcriptional responses illustrated in Figure 3.41.

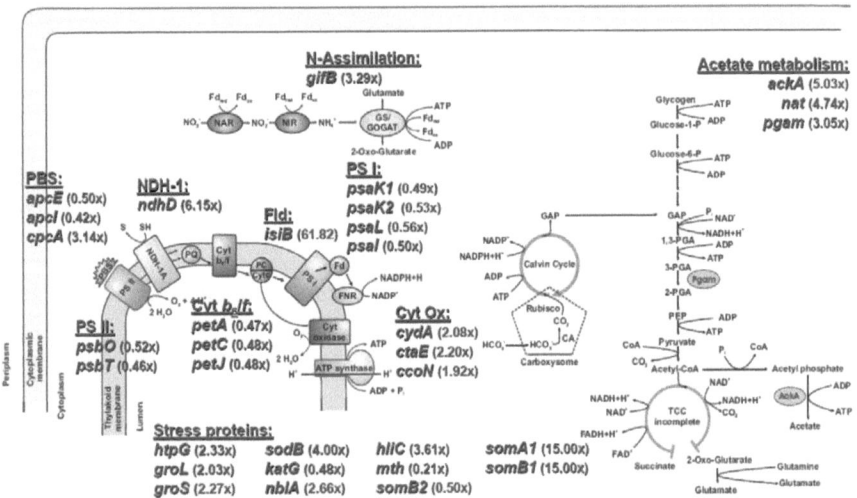

Figure 3.41: Schematic presentation of major transcript changes in *S. elongatus* PCC 7942 WT in the course of iron starvation.

3.7.7 Major differences between *S. elongatus* PCC 7942 WT, mutant K10, and mutant MuD in the acclimation to iron limitation

The *S. elongatus* PCC 7942 mutant K10 lacks IdiB but contains IdiC, while the *idiC*-merodiploid mutant MuD has a strongly reduced amount of IdiC and IdiB (Pietsch et al. 2007). In mutant K10 as well as in mutant MuD, the steady-state concentration of the *idiA* mRNA level as shown previously (Yousef et al. 2003), and the mRNA level for IrpA/IrpB, Som(A2), Som(B1), a Ftr1-similar protein as well as for a putative acetate kinase (AckA), an N-acetyltransferase (Nat), and the phosphoglycerate mutase (Pgam) was substantially lower than in WT. This suggests that the iron-responsive transcriptional activator IdiB regulates the expression of (1) a putative high-affinity iron uptake system consisting of IrpA/B and Som(A2), Som(B1) and possibly Ftr1, (2) the protein IdiA having a function in protecting PS II (Exss-Sonne et al. 2000), (3) enzymes having a function in acetate metabolism (AckA and Nat), and (4) phosphoglycerate mutase (Pgam) suggesting that intermediates of glycogen fermentation and of Calvin cycle such as 3-phosphoglycerate might in part be used for synthesis of acetyl phosphate resulting in an additional substrate-level phosphorylation site for ATP synthesis. An alignment of the putative IdiB-binding sites for the genes *irpA*, *ftr1*, *somB(1)*, *ackA*, and *pgam* is given (see Figure 3.42). The binding of IdiB to the upstream DNA region of *idiA* was previously proven experimentally (Michel et al. 2001), while the interaction of IdiB with the other five upstream DNA regions of the genes await experimental proof.

Figure 3.42: Comparison of the IdiB-binding site in the upstream DNA region of *idiA* with putative IdiB-binding sites upstream of the genes *irpA*, *somB(1)*, *ftr1*, *ackA*, and *pgam* of *S. elongatus* PCC 7942. The binding site of the transcription factor IdiB in the upstream DNA region of *idiA* has been verified experimentally (Michel et al. 2001). The transcripts of the other above mentioned five genes accumulated under iron starvation in WT but not in mutants K10 and MuD, which have either no or only a very low amount of IdiB protein. Putative IdiB-binding sites are boxed in light grey, putative ribosome-binding sites are boxed in black with white lettering, and the annotated start codons are boxed. For *idiA* the transcriptional start site was experimentally determined and is indicated as +1.

Although *nat* and *somA(2)* seem to be also regulated by IdiB, these genes lack the characteristic IdiB-binding site. Figure 3.43 illustrates the deduced consensus sequence of the IdiB binding site.

The above discussed observation implies that in the mutants K10 and MuD the putative high-affinity iron uptake system IrpA/IrpB is not as efficiently working as in WT and that as a consequence, the cells suffer more rapidly from the consequences of iron limitation than that of WT.

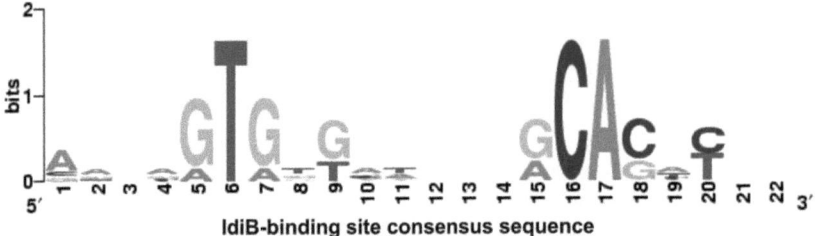

IdiB-binding site consensus sequence

Figure 3.43: IdiB consensus-binding sequence determined by comparison of the nucleotide sequence of IdiA with the putative members of the IdiB regulon. Alignment of the upstream DNA regions containing the putative IdiB-binding site of the genes *idiA*, *irpA*, *irpB*, *somA*, *somB(1)*, *ftr1*, *ackA*, and *pgam* from *S. elongatus* PCC 7942 with the WebLogo tool at http://weblogo.berkeley.edu/. This software displays the similarities of single bases with the size of the corresponding character. The larger the character, the higher is the degree of concordance. Thus, the main elements of the IdiB-binding site are present in the nucleotide sequence of all investigated genes.

Moreover, the lack of IdiA prevents the protection of PS II by this protein and thus, iron limitation causes a faster proceeding damage of PS II. This is in line with the lower O_2 evolving activity observed in the two mutants as compared to WT (Pietsch et al. 2007). The reduced transcript level for enzymes in favor of acetate-phosphate biosynthesis (Pgam and AckA) indicates that the mutant cells can not benefit from an additional site of substrate-level phosphorylation via the conversion of acetyl-phosphate to acetate with concomitant synthesis of ATP to the same extent as WT (Moezelaar et al. 1996; Moezelaar and Stal 1994; Steunou et al. 2006; van der Oost et al. 1989). The latter difference suggests that the strategies to minimize imbalances in the C/N ratio and/or the NADPH/ATP ratio under iron limitation are slightly different in WT as compared to the two mutants due to the absence or reduced concentration of IdiB and IdiC. Presently, the regulatory protein(s) for the *idiCB* operon remain(s) unidentified.

The detected transcript levels for regulatory proteins reveal a major difference between the transcriptome of iron-depleted WT and mutant cells with respect to a transcript for a CheY-like two-component response regulator. The steady-state transcript level was found to be elevated in WT, but was decreased in both mutants. Further, the transcript for the group II σ factor RpoD4 was found at a higher level in WT, while the corresponding transcript level diminished in mutants K10 and MuD grown for 72 h in iron-depleted medium.

Due to the strongly reduced amount of IdiC and the reduced amount of IdiB in MuD, a number of additional changes in transcript abundance relative to WT and mutant K10 were observed for mutant MuD - especially with respect to photosynthesis-related transcripts (see Table 3.3 B). The results suggest that the modification of the electron transport chain due to iron starvation from a preferentially photosynthetic linear transport to a preferentially photosynthetic cyclic and respiratory electron transport in mutant MuD did not proceed equally well in mutant MuD as in WT. This became particularly obvious, when the mRNA levels for *isiA*, *isiB*, and *isiC* were compared in the three strains - suggesting differences in the redox signals mediated by the electron transport chain. This fact might also explain why it has been impossible to obtain a fully-segregated IdiC-free mutant, while it has been possible to obtain a fully-segregated IdiB-free mutant.

3.8 Preliminary work: identification of putative transcription regulators controlling the transcritpion of the *idiB* operon in *S. elongatus* PCC 7942

Even though the proteins IdiA, IdiB, and IdiC are well characterised, the putative regulator for iron-regulated gene expression of the *idiB* operon is so far still unknown. Previous results postulated some kind of a transcriptional regulator which shuts off *idiB* transcription under iron-repleted growth conditions. Moreover, a model has been presented which refers to the observed interaction of *idiB* transcription in the cause of iron starvation as well as in the course of progressing oxidative stress (Yousef et al. 2003). This model suggested the presence of a PerR-like transcriptional master regulator, which has the potential to sense iron starvation as well as oxidative stress. However, previous attempts to localise the putative promotor of the *idiB* operon initiating the transcription of the primary tricistronic transcript showed rather conflicting results (Lim 2003). Although a stable 5'-end of the *idiB* transcript has been mapped with primer extension assays, no corresponding promotor activity was detected for the corresponding *idiB::luxABCDE* constructs in *lux* reporter gene assays. Thus, another major aim of this work was to identify a putative transcriptional regulator, which regulates the iron-dependent gene expression of the *idiB* operon.

3.8.1 Identification of a novel Fur-homologous transcriptional regulator in *S. elongatus* PCC 7942

The Fur protein had been shown to negatively regulate the expression of *isiA* in *S. elongatus* PCC 7942 (Ghassemian and Straus 1996). In the presence of its co-repressor Fe^{2+}, Fur binds in a dimeric state to specific operator sequences, called Fur-boxes, upstream of the corresponding genes and represses their transcription. Moreover, preliminary work had already postulated the presence of an additional PerR-like transcriptional regulator as a candidate for an iron-dependent and oxidative stress-responsive transcriptional regulator in *S. elongatus* PCC 7942 and in particular, for the *idiB* operon (Yousef et al. 2003). The peroxide regulon repressor PerR is a metalloprotein homologous to Fur. The regulation of PerR activity involves both metal ions and oxidation of cysteines (Fuangthong et al. 2002). A dual control of expression in response to iron and oxidative stress has e.g. also been demonstrated for the expression of the *fldA* gene encoding flavodoxin in *E. coli*. Its expression is regulated by two transcription factors, one is Fur and the other is SoxS (Zheng et al. 1999). Another example is the regulation of catecholate siderophore biosynthesis in *Azotobacter vinelandii* being repressed by a Fur-Fe^{2+}-iron complex and activated by another DNA-binding protein in response to superoxide stress (Tindale et al. 2000).

Knowing that at least PerR from *Synechocystis* sp. PCC 6803 has recently been shown to regulate a distinct regulon different from the Fur regulon (Li et al. 2004), the genome of *S. elongatus* PCC 7942 has been screened for candidate genes encoding PerR/Fur-like transcriptional repressors, which might represent candidate genes to encode the transcriptional regulator of the *idiB* operon. Blast searches with selected genes identified the JGI open reading frame gene 2170 as one such candidate. The corresponding gene product contains 152 amino acid residues, has a calculated molecular mass of 17.3 kDa, and has a pI of 7.74 (ProtParam, ExPASy). This finding fits previous reports suggesting that several cyanobacterial strains contain more than one *fur*-homologous gene in their genomes (Kunert et al. 2003).

An alignment of the putative amino acid sequence of FurII and the Fur repressor from *S. elongatus* PCC 7942 was performed using the ClustalW software at EMBL (see Figure 3.44). The amino acid similarity between both proteins corresponds to 61% similar amino acid residues including 42% identical amino acid residues.

```
FurII    VPTVTVSSLSAVRVAALRSALEQAGYRLTPQRYWIAEIFEGLAQGEHLSAIDLQRHLSDR  60
Fur      --------MTYTAASLKAELNERGWRLTPQREEILRVFQNLPAGEHLSAEDLYNHLLSR  51
                 .*:*::  *::  *;*******  *  .:*:.*.  ****** **  **  .*

FurII    QTPLSKSTIYRSLESLCHAAWLRCITLDRKQRCYELNR-EGTHYHLTCLHCQAVIEFLDD 119
Fur      NSPISLSTIYRTLKLMARMGLLRELDLAEDHKHYELNQPLKHHHHLICVSCSKTIEFKSD 111
         ::*:* *****:*: :.: . ** : * ,.:: ****:     *:** *: *. .*** .*

FurII    RVIHLSEGIADRYGFQLLNCQLLIMGICAACRY--- 152
Fur      SVLKIGAKTSEKEGYHLLDCQLTIHGVCPTCQRSLV 147
         *::::.   ::: *::**:*** * *:*.:*:
```

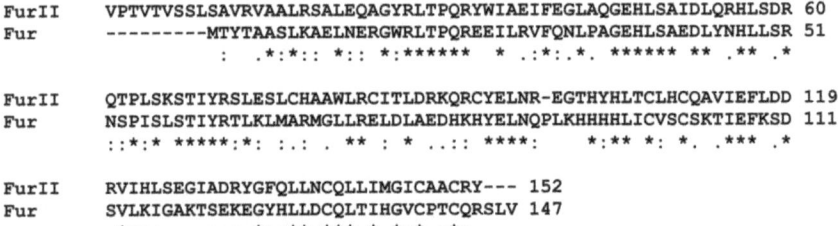

Figure 3.44: Alignment of the transcriptional repressor Fur and the Fur-homologue FurII from *S. elongatus* PCC 7942 using the ClustalW software at EMBL (http://www.ebi.ac.uk/clustalw/). Amino acid similarity corresponds to 61% similar and 42% identical amino acid residues. FurII has a total length of 152 amino acid residues (17.3 kDa) and Fur has a total length of 147 amino acid residues (16.8 kDa). * indicate identical amino acid residues, : indicate conservative substitutions (A/V/F/P/M/I/L/W, D/E, R/H/K, S/T/Y/H/C/N/G/Q, and X indicate mostly conserved amino acid residues. Gaps were introduced into the sequences to maintain an optimal alignment.

Quite surprisingly, the *furII* gene is located immediately downstream of the *dpsA* gene, which is located next to the *idiB* operon (see Figure 3.45). To investigate whether the Fur-homologous protein FurII from *S. elongatus* PCC 7942 is indeed responsible for the iron-dependent transcription of the *idiB* operon, a *furII*-insertionally inactivated *S. elongatus* PCC 7942 mutant was constructed. The *furII* gene sequence including an extra 1,000 bps up and downstream of *furII* was PCR-amplified using gene-specific primers (see Table 2.4). The amplified PCR product was cloned in a *Sma*I-cleaved and dephosphorylated pUC19 vector (NEB). Further on, the *furII* allele was insertionally inactivated by the insertion of a spectinomycin resistance cassette obtained from plasmid pHP45Ω (Fellay et al. 1987) and flanked by Ω terminators preventing any read-through transcriptional activity into a single *Xho*I site of the *furII* sequence. A physical map of the *furII* gene region is shown in Figure 3.45.

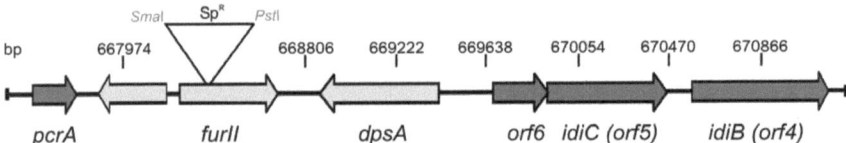

Figure 3.45: Physical map of the chromosomal region surrounding the *furII* gene in *S. elongatus* PCC 7942. The gene *furII* is located immediately upstream of *dpsA*.

Following the transformation assay, four ampicillin-sensitive and spectinomycin-resistant *S. elongatus* PCC 7942 clones were tested showing fully segregration of the *furII*-mutant allele (not shown). A clone named K9#1 was shown to be fully segregated and chosen for all subsequent investigations (see Figure 3.46).

Comparative growth measurement of mutant K9#1 and *S. elongatus* PCC 7942 WT revealed a cell appearance of the mutant that was quite different from that of WT. Although the *furII* insertionally inactivated mutant showed an almost identical growth rate as compared to WT under iron-sufficient as well as under iron-deficient growth conditions, the pigmentation was found to be quite different and resembled that of the IdiB-free mutant K10 (see chapter 3.6).

As illustrated in Figure 3.47, mutant K9#1 contained much lower amounts of Chl as compared to WT. This reduction in Chl content was strongly pronounced under iron-deficient growth conditions. Measurements to characterise the phenotype of the mutant in greater detail remain to be done.

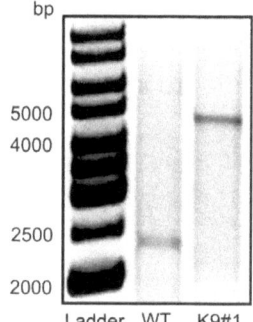

Figure 3.46: Colony PCR of *S. elongatus* PCC 7942 WT and the *furII*-insertionally inactivated mutant K9#1 using primers for the amplification of the *furII* gene including 1,000 bp up and downstream, respectively. The WT showed the PCR product including the WT allele, whereas mutant K9#1 contained only the mutated allele including the spectinomycin resistance.

To investigate whether FurII is indeed involved in iron-dependent transcription of the *idiB* operon, *S. elongatus* PCC 7942 WT and the *furII* knock-out mutant K9#1 were inoculated with a cell density corresponding to an optical density at 750 nm of 0.4 and cultivated with regular BG11 medium and with BG11 medium from which iron was omitted. Cells were grown for 96 h, harvested, and subsequently used for the preparation of cell-free extracts.

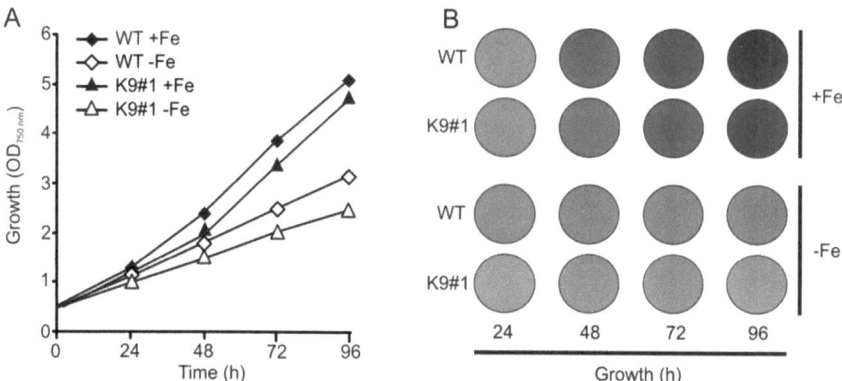

Figure 3.47: Growth curves and appearance of *S. elongatus* PCC 7942 WT and the FurII-free mutant K9#1 when grown under iron-sufficient or iron-deficient conditions. **(A)** *S. elongatus* PCC 7942 WT and mutant K9#1 were grown either in iron-sufficient or iron-deficient BG11 medium for 96 h. ♦ WT in iron-sufficient BG11 medium; ◊ WT in iron-deficient BG11 medium; ▲ mutant K9#1 in iron-sufficient BG11 medium; △ mutant K9#1 in iron-deficient BG11 medium. **(B)** Appearance of WT and mutant K9#1 cells from the experiments as shown under (A). Under iron-deficient growth, cells of mutant K9#1 had a lower pigment content than WT. The cells appeared after 96 h in a bleached yellow colour. Under regular growth conditions K9#1 showed only in the early growth phase a slightly decreased pigmentation compared to that of WT.

Immunoblot analyses with cell-free extracts from WT and the FurII-free mutant K9#1 using antisera against IdiB, IdiA, IrpA, and IsiA revealed that none of the tested proteins was substantially up- or down-regulated in the mutant as compared to WT under iron-sufficient or iron-deficient growth conditions (Figure 3.48). This result clearly excludes a function of the FurII protein from *S. elongatus* PCC 7942 as the transcriptional regulator of the *idiB* operon.

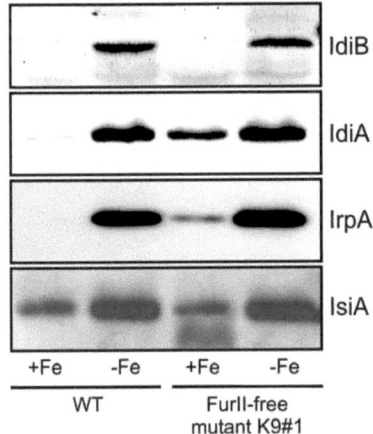

Figure 3.48: Immunoblot analysis of the expression of IdiB, the IdiB-regulated proteins IdiA and IrpA as well as the Fur-regulated protein IsiA. Cells were harvested after 96 h of growth with iron-sufficient or iron-deficient BG11 medium. Cell-free extracts corresponding to 50 µg protein were subjected to SDS PAGE and immunoblotting. The used antisera were diluted as follows: anti-IdiB 1:500; anti-IdiA 1:2,000; anti-IrpA 1:750; anti-IsiA 1:500. Detection was performed using ECL™ detection kit (GE Healthcare).

These results were also confirmed by preliminary results of a DNA microarray experiment comparing the transcriptomes of WT, a *fur*-merodiploid mutant, and the FurII-free mutant K9#1. Cells were grown under iron-sufficient and iron-deficient conditions harvested after 72 h of growth for extraction of total RNA. In the here presented DNA microarray experiments, the iron-deficiency transcriptomes of the three strains were compared to the iron-sufficient transcriptome of the identical strain and the iron-sufficient transcriptome of mutant K9#1 was compared to the iron-sufficient transcriptome of *S. elongatus* PCC 7942 WT. The results for the transcripts of selected major iron-regulated genes are given in Table 3.7. Since no significant differences in transcript levels were detected, FurII seems to be involved neither in regulation of the genes belonging to the IdiB regulon nor in the regulation of the Fur regulon under iron-deficient growth conditions.

Table 3.7: List of transcriptionally-regulated genes in S. elongatus PCC 7942 WT, the fur-merodiploid mutant, and the FurII-free mutant in response to growth for 72 h in the presence or absence of iron in BG11 medium. The table contains the evaluated data of a single biological experiment without a dye-swap experiment. The fold change value is calculated as $\log_2^{\text{M-value}}$ of M-values with corresponding p-value ≤0.051. M-values >-0.90 and <+0.90 indicate no significant change in the transcriptional levels (corresponding to a fold change of ≤1.87 and ≥0.53). Significantly increased or decreased transcript levels are printed in bold letters. JGI ORFs correspond to the JGI annotation. Common gene names are given in the second column. "Error" software related failure.

JGI ORF	Gene	Annotated protein function	Growth for 72 h -Fe vs. +Fe Fold change of transcripts		
			WT	Fur	FurII
1462	irpA	Iron-regulated protein A	13.91	6.10	10.48
1461	irpB	Multiheme c-type cytochrome family protein with two heme-binding sites	18.40	6.30	12.65
1463	somB(1)	Major outer membrane protein probably forming porin-like β-barrel structure and which might also connect to the S-layer	2.22	2.38	1.53
1464	somA(1)	Major outer membrane protein probably forming porin-like β-barrel structure and which might also connect to the S-layer	2.17	error	error
1607	somA(2)	Major outer membrane protein, see above	3.41	2.85	2.50
1407	futB	Iron (III) ABC transporter permease	error	error	2.71
1733	sufR	Repressor of the suf regulon	2.27	error	2.79
1736	sufC	[Fe-S]-assembly ATPase SufC	2.36	error	2.42
1738	sufS	Cysteine desulfatase, NifS-similar, involved in formation of [Fe-S] centres	error	error	2.33
1541	isiB	Flavodoxin, soluble electron transport protein, in part replaces ferredoxin under iron starvation	17.26	5.28	14.42
1540	isiC	Putative hydrolase with typical αβ-fold of hydrolases	30.06	4.59	22.31
2175	idiA	Iron deficiency-induced protein A, modifies and protects photosystem II against selected stresses	3.90	4.38	2.30
2174	idiB	Iron deficiency-induced protein B, positively acting transcription factor of IdiA and the IdiB regulon	7.11	2.98	3.78
2173	idiC	Iron deficiency-induced protein C, suggested to participate in photosynthetic cyclic electron transport	14.92	12.30	10.41
2172	orf6	Gene immediately upstream of idiC, encodes a protein of unknown function	10.00	5.46	7.57

Results

Although the presented immunoblots exclude the transcriptional regulation of the *idiB* operon by FurII, they reveal another interesting result. As can be seen, the mutant K9#1 contains substantially higher amounts of IdiA and IrpA under regular growth conditions as compared to WT. Since this finding might be indicative for some kind of elevated oxidative stress in the mutant in comparison to WT and since it has been described in the literature that Fur-homologous proteins may have a function in the transcriptional response to oxidative stress, this result may be related to a function of FurII in the adaptational process to oxidative stress.

In order to examine a possible function of FurII in sensing oxidative stress, cultures of *S. elongatus* PCC 7942 WT and the FurII-free mutant K9#1 were inoculated with a cell density corresponding to an optical density at 750 nm of 0.4 and cultivated for 24 h under regular growth conditions. To apply oxidative stress, the cultures were treated with H_2O_2 to give a final concentration of 2 mM, 5 mM, 8 mM, and 10 mM.

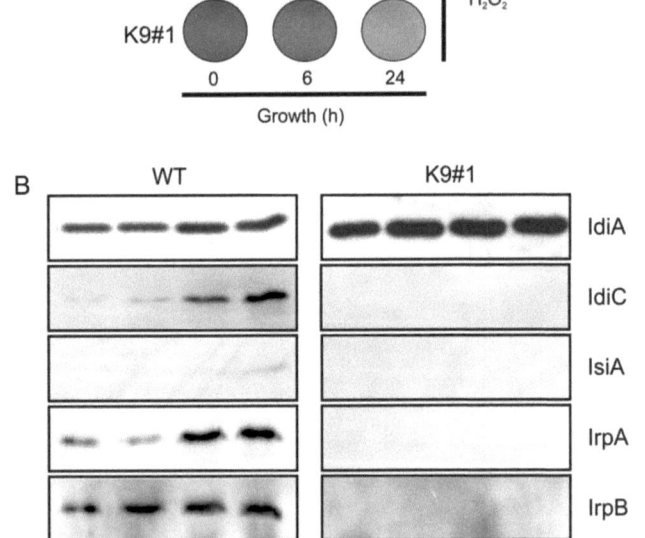

Figure 3.49: Investigation of growth and protein expression of the FurII-free mutant K9#1 after addition of H_2O_2 causing oxidative stress. **(A)** Cell appearance of mutant K9#1 and of *S. elongatus* PCC 7942 WT 6 h and 24 h after addition of 10 mM H_2O_2. In contrast to WT, mutant cells showed a highly H_2O_2-sensitive phenotype. **(B)** Immunoblot analyses of mutant K9#1 and *S. elongatus* PCC 7942 WT after addition of 2 mM, 5 mM, and 8 mM H_2O_2. Cells were harvested 6 h after treatment with H_2O_2. Cell-free extracts corresponding to 100 µg protein were subjected to SDS PAGE and immunoblotting. Immunoblots were performed with the anti-IdiA (dilution 1:1,000), anti-IdiC (dilution 1:300), anti-IsiA (dilution 1:1,000), anti-IrpA (dilution 1:500), anti-IrpB (1:1,000), and the anti-DpsA antiserum (dilution 1:2,000) and the ECL™ detection kit (GE Healthcare).

Growth, cell appearance, and the expression of selected proteins were investigated after growth for additional 6 and 24 h. The results clearly showed that the mutant was strongly impaired in its cellular response to increased oxidative stress triggered by the addition of H_2O_2. As illustrated in Figure 3.49 A, the mutant responded strongly to the addition of 10 mM H_2O_2. Within 6 h after the addition of H_2O_2, the mutant culture showed a strong bleached phenotype. After 24 h the cell culture showed a blue colour, which indicates cell lysis and cell death. On the contrary, WT cells showed just a slightly bleached phenotype and seemed to have fully recovered from the treatment after 24 h.

In order to investigate the response of mutant K9#1 to oxidative stress in greater detail, WT and mutant cells were exposed to gradually increasing forms of oxidative stress in the presence of 2 mM, 5 mM, and 8 mM H_2O_2 (see Figure 3.49). The corresponding cell-free extracts of WT and the mutant were applied to immunoblotting with antisera against IdiA, IdiC, IsiA, IrpA, IrpB, and DpsA. Quite intriguingly, the mutant showed no detectable amounts of IdiC, IsiA, IrpA, and IrpB under any tested concentration of hydrogen peroxide whereas the WT showed a regular stress response. In contrast to the absence of all these proteins, the mutant contained high levels of IdiA which exceeds the IdiA concentrations in WT under oxidative stress. Although these preliminary results give new insights in the transcriptional regulation of the oxidative stress response in *S. elongatus* PCC 7942, it is by far too early to draw final conclusions concerning a putative role of FurII as a master regulator of the oxidative stress response. Since these experiments have been performed only once, they have to be repeated again and have to be extended and complemented by DNA microarray experiments.

3.8.2 Identification of a novel MerR-like transcriptional regulator in *S. elongatus* PCC 7942

In addition to the Fur-homologous transcriptional regulator FurII, the results obtained from the microarray experiments shown in chapter 3.7 led to the identification of a second novel iron starvation induced regulatory gene in *S. elongatus* PCC 7942. The gene *merR* encodes a putative oxidative stress-sensing transcriptional regulator. This type of regulator was at first described in *E. coli* as a transcriptional activator of the *mer* genes in the presence of Hg(II) salts and as a weak repressor in the absence of Hg(II) (Lund and Brown 1989; Lund et al. 1986). In addition, MerR also regulated its own transcription. MerR belongs to the family of HTH transcription factors including SoxR of *E. coli*. SoxR e.g. activates the transcription of flavodoxin (Brown et al. 2003). The MerR family of transcriptional regulators form dimeric proteins that display homologous N-terminal DNA-binding domains, which are linked by variable length coiled loops to ligand-specific C-terminal "coactivator" binding domains (Newberry and Brennan 2004). These transcriptional regulators are activated in response to stress signals in eubacteria, such as exposure to oxygen radicals, heavy metals, or cytotoxic compounds (Brown et al. 2003). MerR regulators sense peroxides by metal-catalysed oxidation and therefore, contain redox active [4Fe-4S]-centres. Protein oxidation, catalysed by the bound iron cofactor, leads to the rapid conformational change of the protein. This mechanism accounts for the ability of MerR to sense low levels of H_2O_2 *in vivo*. The MerR family members activate transcription from σ^{70}-targeted promoters in Gram-negative bacteria and σ^A/ σ^{54}-targeted promoters in Gram-positive bacteria that contain suboptimal 19-bp spacers between their -35 and -10 promoter elements, which results in the misalignment of these promoter elements and preclusion of an open complex formation by RNA polymerase (Ansari et al. 1995; Outten et al. 1999).

Thus, the *merR* gene (JGI open reading frame 1739) in *S. elongatus* PCC 7942 represents a second good candidate to encode the transcriptional regulator of the *idiB* operon. The corresponding gene product 1739 codes for a protein of 141 amino acid residues in length, has a calculated molecular mass of 16.1 kDa, and has a pI of 8.93 (ProtParam, ExPASy). The deduced amino acid sequence of MerR shows high similarity to different members of the MerR family (BlastP, NCBI; Pfam Motif Search, Sanger).

An alignment of the putative amino acid sequence of *merR* from *S. elongatus* PCC 7942 and the MerR regulator of *Cyanothece* sp. PCC 8801 was performed using the ClustalW software at EMBL and is illustrated in Figure 3.50. The amino acid similarity between both proteins corresponds to 74% similar amino acid residues including 54% identical amino acid residues.

```
MerR S. elongatus PCC 7942      --LTATLLKIGEIAKQVGVAVGTIRYYETLQLIQPSTRGENGYRYYKPQT  48
MerR Cyanothece sp. PCC 8801    MMNTSSYLKIGELAQQTGLSVGNLRYYSDLGLLEPVTRGENGYRYYSQQA  50
                                  *::  *****:*:*.*::**.:***. * *::*  **********. *:

MerR S. elongatus PCC 7942      IQQLQFIRQAQTLGFSLEEIRQILTVYAEGTPPCSLVQTLLNQKIATLEE  98
MerR Cyanothece sp. PCC 8801    TQQVEFIKKAQAIGFTLEEIKQILDVRDRGETPCHLVQTLLDHKIEELEI 100
                                 **::**::.**::**:****:***  *   .*  .** ******::** **

MerR S. elongatus PCC 7942      KLQQIQTFKAQLESYRDRWQQTPTPQTVTKSEICPLIATIPQT------ 141
MerR Cyanothece sp. PCC 8801    KIKQMTLFKSELEGYRTDWIIHPHLQSSS-AEICPLISSVSLNSEHNTP 148
                                 *::*: **::**.** *   *:  :******::.. .
```

Figure 3.50: Alignment of the deduced amino acid sequence of *merR* from *S. elongatus* PCC 7942 and the MerR regulator from *Cyanothece* sp. PCC 8801 using the ClustalW software at EMBL (http://www.ebi.ac.uk/clustalw/). Amino acid similarity corresponds to 74% similar and 54% identical amino acid residues. MerR of *S. elongatus* PCC 7942 has a total length of 141 amino acid residues (16.1 kDa) and MerR of *Cyanothece* sp. PCC 8801 has a total length of 148 amino acid residues (16.9 kDa). * indicate identical amino acid residues, : indicate conservative substitutions (A/V/F/P/M/I/L/W, D/E, R/H/K, S/T/Y/H/C/N/G/Q, and X indicate mostly conserved amino acid residues. Gaps were introduced into the sequences to maintain an optimal alignment.

Interestingly, the *merR* gene of *S. elongatus* PCC 7942 belongs to the iron-dependent regulated *suf* gene cluster and is located immediately downstream of *sufS* (see chapter 3.7.1 and Figure 3.51). In order to investigate whether MerR of *S. elongatus* PCC 7942 is involved in the iron-dependent transcription of the *idiB* operon, the plasmid 8S4-G1 was obtained from Prof. PhD Susan Golden (Texas A&M Universitiy, College Station Texas, USA) and used for the inactivation of *merR* in *S. elongatus* PCC 7942 via transposon mutagenesis. This construct was a derivative of pMCL200 with ~8 kb of genomic DNA and 1.2 kb of extra transposon DNA carrying a kanamycin resistance in the transposon and chloramphenicol resistance in the vector. This construct was amplified and transformed in *S. elongatus* PCC 7942 WT cells. Four chloramphenicol-sensitive and kanamycin-resistant clones were tested and showed full segregration of the *merR*-mutant allele (not shown). A clone named 2#1 was chosen for subsequent experiments.

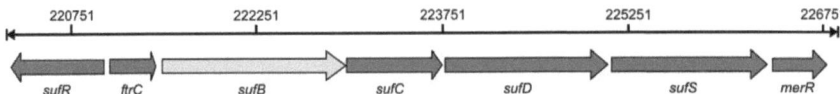

Figure 3.51: Physical map of the chromosomal region containing the *suf* gene cluster including the *merR* gene in *S. elongatus* PCC 7942. The gene *merR* is located immediately downstream of *sufS*.

In order to examine whether MerR is the transcriptional regulator of the *idiB* operon, cells of the MerR-free mutant 2#1 were shifted from BG11 agar plates to liquid medium. After 96 h of growth with regular BG11 medium, it became obvious that the putative MerR transcription regulator in *S. elongatus* PCC 7942 must represent the superior regulator for gene regulation even under regular growth conditions. In contrast to WT, the mutant cells showed no longer a significant growth after an initial growth phase for 48 h and were almost completely bleached after 96 h (see Figure 3.52). The corresponding BG11 medium revealed a strongly decreased pH value of 6.3 indicating that the mutant cells were not able to maintain a physiological pH during growth in regular BG11 medium. When the same cultures were grown in the presence of 20 mM HEPES pH 7.5 and 20 mM NaHCO$_3$, the cells recovered within one week. The growth of mutant 2#1 with regular BG11 medium and BG11 medium containing 20 mM HEPES pH 7.5 for 96 h, as compared to that of *S. elongatus* PCC 7942 WT, is illustrated in Figure 3.52.

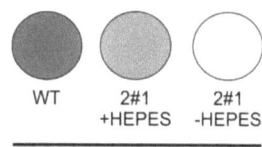

Figure 3.52: Growth of the MerR-free *S. elongatus* PCC 7942 mutant 2#1 in regular BG11 medium without HEPES buffer and in BG11 medium containing 20 mM HEPES pH 7.5 for 96 h compared to WT grown in regular BG11 medium. The result showed that the MerR-free mutant lost its ability to maintain a physiological pH value in the medium.

Cells of mutant 2#1, which were grown with BG11 medium containing 20 mM HEPES pH 7.5, revealed a constant, but strongly reduced growth as compared to WT grown in the same medium (see Figure 3.53 A). In contrast to the FurII-free mutant K9#1, the growth of mutant 2#1 was highly reduced even under regular growth conditions. After 96 h of growth in regular BG11 medium, mutant 2#1 contained only a Chl *a* content of 23.5% as compared to that of WT. On the contrary, the Chl *a* content of iron-starved 2#1 cells was reduced about 75% as compared to that of WT (see Figure 3.53 B). In fact, the growth of mutant cells grown with iron-deficient BG11 medium was reduced compared to those grown under regular conditions, but compared to WT the decrease in growth was higher under iron-sufficient conditions. Thus, the loss of MerR seems to result in major reduction of the viability of *S. elongatus* PCC 7942 under the used conditions.

It is obvious that the MerR regulator of *S. elongatus* PCC 7942 plays a superior role in gene regulation even under regular growth conditions. Immunoblot analysis with the anti-IdiA antiserum using cell-free extracts obtained from mutant 2#1 cells grown for 96 h with iron-deficient BG11 medium, which contained 20 mM HEPES pH 7.5, showed a slightly reduced IdiA level as compared to that of WT. Since IdiB is the positively acting transcription factor of IdiA, the MerR regulator certainly does not represent the main transcriptional regulator of the *idiB* operon. In contrast to the FurII-free mutant K9#1, mutant 2#1 showed only minor detectable IdiA amounts under iron-sufficient conditions. Thus, the reduced growth and pigment content of this mutant could not be due to oxidative stress.

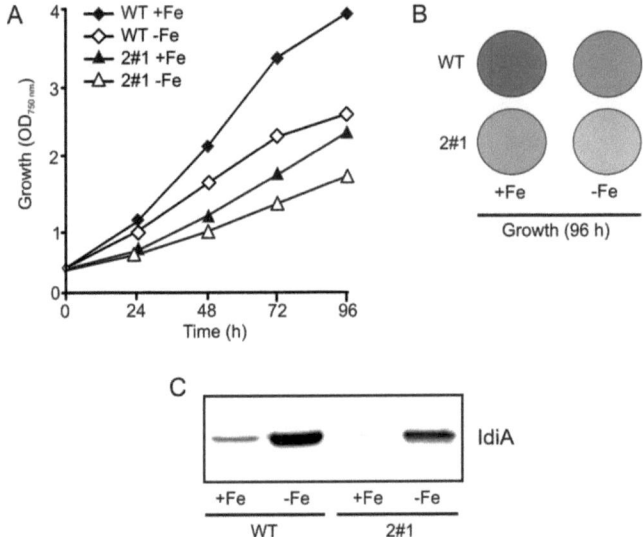

Figure 3.53: Growth, cell appearance, and IdiA expression of the MerR-free mutant 2#1 and *S. elongatus* PCC 7942 WT when grown with iron-sufficient or iron-deficient BG11 medium containing 20 mM HEPES pH 7.5. Cells were inoculated with an optical density at 750 nm of 0.4. **(A)** *S. elongatus* PCC 7942 WT and mutant 2#1 were grown either in iron-sufficient or iron-deficient BG11 medium with 20 mM HEPES pH 7.5 for 96 h. ♦ WT in iron-sufficient BG11 medium; ◊ WT in iron-deficient BG11 medium; ▲ mutant 2#1 in iron-sufficient BG11 medium; △ mutant 2#1 in iron-deficient BG11 medium. **(B)** Appearance of WT and mutant 2#1 cells harvested after 96 h of growth as described in (A). Mutant 2#1 showed a highly reduced growth and a decreased pigmentation compared to that of WT even under regular growth conditions. Under iron-deficient growth, cells of mutant 2#1 had also a lower pigment content than WT and appeared after 96 h in a bleached yellow colour. **(C)** Immunoblot analyses of mutant 2#1 grown for 96 h with iron-sufficient and iron-deficient BG11 medium containing 20 mM HEPES pH 7.5. Cell-free extracts corresponding to 50 µg protein were subjected to SDS PAGE and immunoblotting. Immunoblots were performed with the anti-IdiA antiserum (dilution 1:1,000) and the ECL™ detection kit (GE Healthcare).

However, in contrast to the MerR regulators described in bacteria, these preliminary results suggest that the MerR regulator is not involved in the transcriptional regulation of the oxidative stress response in *S. elongatus* PCC 7942. In fact, it seems to play a major role in superior gene regulation independent of any stress conditions. Since the *merR* gene belongs to the *suf* operon, it is assumed to be involved in the regulation of the *suf* genes in addition to SufR in *S. elongatus* PCC 7942. If the Suf system plays a role in addition or instead the Isc system as the house-keeping [Fe-S] assembly system in *S. elongatus* PCC 7942, the lack or major reduced amount of functional [Fe-S] centres could explain the extreme phenotype of mutant 2#1 during growth with regular BG11 medium.

4 Discussion

During the last decade, more and more experimental evidence has accumulated suggesting that iron starvation is a specific form of a micro nutrient limitation, which has the potential to severely limit biomass production of many organisms, as e.g. oxygenic photosynthetic organisms in otherwise nutrient-rich habitats (Behrenfeld and Kolber 1999; Geider and La Roche 1994; Martin et al. 1994; Tortell et al. 1999). This is especially true for marine cyanobacterial species in vast regions of the world's oceans, which are considered to contribute up to 40% of the global biomass production and also to contribute to global nitrogen fixation (Paerl 2000).

Despite its hazardous potential, iron is involved in a large number of cyanobacterial metabolic key reactions of photosynthesis, respiration, nitrogen assimilation, sulfur assimilation, and DNA synthesis (Boyer et al. 1987; Straus 1994). In particular, the photosynthetic electron transport chain and also the respiratory chain contain a high number of iron-containing proteins (Figure 4.1). Thus, iron starvation has a large impact on the function of the electron transport chains. This is especially important, since damaged photosynthetic electron transport chain reactions are known to strongly increase the formation of ROS as well (Aro et al. 1993; Asada 1994; Mittler 2002).

In spite of their long evolutionary history and the fact that cyanobacteria were among the first organisms being capable of an oxygenic-type of photosynthesis (Vermaas 2001), cyanobacteria belong to those organisms, which had to face the "iron problem" from the very beginning of the development of an aerobic atmosphere. During their long evolutionary history, cyanobacteria have developed sophisticated mechanisms to overcome or at least to alleviate the effects of iron starvation. The adaptational response of cyanobacteria to iron starvation has been classified into three different categories, acquisition, compensation, and retrenchment (Straus 1994).

For *S. elongatus* PCC 7942, previous work of our and other groups showed that mainly two proteins contributed to the remodelling process of the electron transport chain in response to iron limitation, IdiA and IsiA (Michel and Pistorius 2004). In summary, the results on IdiA and IsiA have shown that under iron limitation IdiA becomes a part of the acceptor side of PS II and protects it from oxidative stress (Lax et al. 2007; Michel and Pistorius 2004), and that IsiA forms a new antenna around PS I trimers (Boekema et al. 2001). Whereas the transcriptional regulation of *idiA* is mediated by the transcriptional activator IdiB (Michel et al. 2001), the transcriptional regulation of *isiA* is facilitated via the transcriptional repressor Fur (Ghassemian and Straus 1996) (Figure 4.1).

The physiological consequences of the adaptation to iron starvation have indicated that the photosynthetic linear electron transport is reduced but that the activity of the photosynthetic cyclic as well as the respiratory electron transport are increased.

Discussion

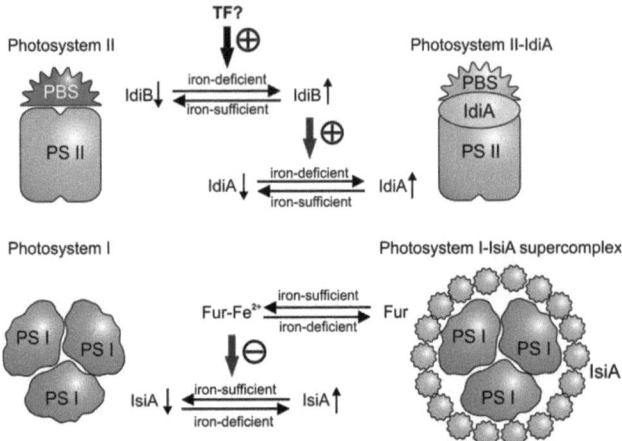

Figure 4.1: Model of the modification of PS II and PS I by IdiA and IsiA in *S. elongatus* PCC 7942 under iron starvation (Michel and Pistorius 2004). Abbreviations are given in the previous legends.

Although the previous work identified and characterised some of the major players of the remodelling process of the electron transport chain in *S. elongatus* PCC 7942 quite well, many questions have remained unanswered. In my work, it was intended to obtain an overall view on the adaptational process to iron limitation by transcript profiling of *S. elongatus* PCC 7942. The second major goal was to gain more information on the function of IdiC in this process.

4.1 Transcript-profiling of *S. elongatus* PCC 7942 and selected mutants grown under iron-deficient and iron-sufficient conditions

To obtain a profound view on the complex regulatory network involved in acclimation to iron limitation in *S. elongatus* PCC 7942 WT and to discover yet unidentified gene products being expressed under iron starvation facilitating the adaptation of the metabolism to the new requirements, DNA microarray analyses with cells grown in the presence or absence of iron in BG11 medium were performed. In addition, two *Synechococcus* mutant strains were included in the investigation. These were an IdiB-free *S. elongatus* PCC 7942 mutant to unravel putative novel members of a concise IdiB regulon and an *idiC*-merodiploid mutant to gain further information on the physiological role of IdiC. This *idiC*-insertionally inactivated mutant strain never showed full segregation implying an essential function of IdiC for the viability of *S. elongatus* PCC 7942. In *S. elongatus* PCC 7942 WT, IdiA, IdiB, and IdiC are up-regulated under iron starvation. In the IdiB-free mutant, IdiB is completely missing, and since IdiB is the transcriptional activator of IdiA expression, the mutant also lacks any IdiA protein. In the *idiC*-merodiploid mutant, at least small amounts of IdiA, IdiB, and IdiC relative to WT are present under iron-limited growth conditions (Figure 4.2 B). These results reflect the gene arrangement of the *idiB* operon illustrated in Figure 4.2 A. Since the gene *idiC* is located directly upstream of *idiB* and the operon is first transcribed into a primary transcript

(Yousef et al. 2003), hardly any IdiB protein could be detected in MuD. In contrast, an effective *idiC* transcription is still possible, given that *idiB* was inactivated but generating reduced amounts of IdiC expression in iron-starved mutant K10 cells as compared to WT.

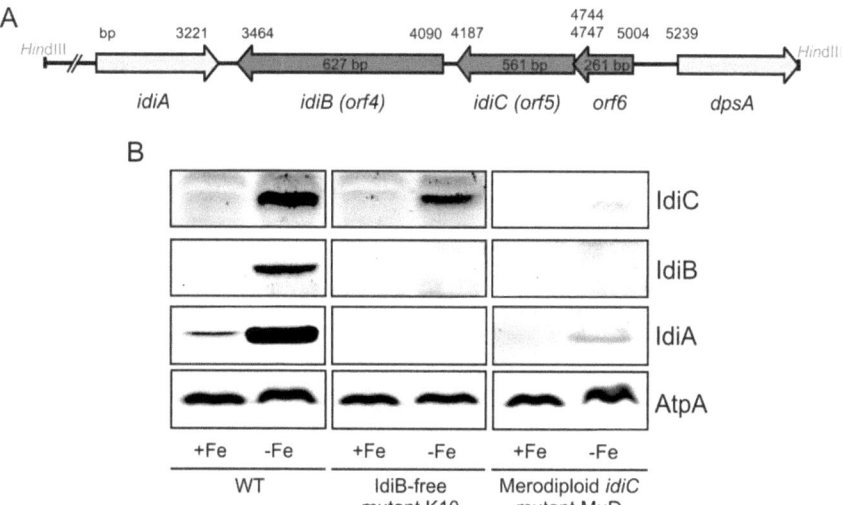

Figure 4.2: (A) Physical map of EMBL database entry Z48754 including the gene idiA and the *idiB* operon. (B) Immunoblot analyses of IdiC, IdiB, and IdiA protein expression in *S. elongatus* PCC 7942 WT, a fully segregated IdiB-free *S. elongatus* PCC 7942 mutant K10, and the *idiC*-merodiploid mutant MuD (Michel et al., 1999). All strains were grown for 72 h under iron-sufficient or iron-deficient conditions.

In summary, the DNA microarray analyses identified six regions on the *S. elongatus* PCC 7942 chromosome with clusters of genes, whose transcripts increased in the course of iron-limited growth conditions:

1. The *irp* region consisting of the genes *irpA*, *irpB*, and *somB(1)*.
2. The *fut* gene region consisting of the genes *futA*, *futB*, and *mapA*.
3. The *suf* gene region consisting of the genes *sufB*, *sufC*, *sufD*, *sufR*, and *sufS*.
4. The *isi* gene region consisting of the genes *isiA*, *isiB*, and *isiC*.
5. The *idi* gene region consisting of the genes *idiA*, *idiB*, and *idiC*.
6. The acetate metabolism region including the genes *ackA* and *pgam*.

Considering the facts, the conclusion has to be drawn that the adaptational response of *S. elongatus* PCC 7942 to iron starvation is somewhat rather "simple", but in evolutionary terms doubtlessly very efficient and "fail proof".

First of all, *S. elongatus* PCC 7942 maximizes its iron acquisition capacity by the expression of the Fut "standard" iron uptake system, which is also found in many other cyanobacterial strains, e.g. in *Synechocystis* sp. PCC 6803 (Katoh et al. 2001a; Katoh et al. 2001b) (Figure 4.5). Since *mapA* lies next to these genes, and since it encodes a protein, which in part resembles solute-binding periplasmic proteins, MapA may also be part of this system.

Moreover, the Fut system is amended by the expression of a novel putative iron uptake system, which consists of the proteins IrpA and IrpB. As already considered for the Fut system, MapA may very well be the solute-binding protein of this system. The fact that IrpB strongly resembles multiheme cytochrome c-type proteins with two CXXC-heme binding sites is quite intriguing, since the enhanced expression of a heme-containing protein and thus, iron-containing protein under iron starvation is puzzling *per se*. However, the presence and nature of IrpB may explain how the transport of iron compounds via the cytoplasmic membrane is energised. Such a system has not been identified yet in any other of the so far sequenced and annotated cyanobacterial strains.

In addition to the Fut and the Irp system, another novel iron transport system seems to complement the gross capability of *S. elongatus* PCC 7942 for enhanced iron acquisition. The Ftr1 system (Figure 4.3) has first been identified in yeast and constitutes together with FET3, a cytoplasmic membrane-located permease for high-affinity iron uptake (Larrondo et al. 2007; Stearman et al. 1996). FET3 is a glycoprotein with copper as metal ion cofactor, which is probably located at the outside of the cytoplasmic membrane. It has been shown that it contains a ferrooxidase activity, which suggests that it efficiently oxidises Fe(II) to Fe(III) for its subsequent transport into the intracellular compartment by the iron permease Ftr1. The cyanobacterial Ftr1 has four typical EXXE motifs as being present in yeast Ftr1, and has a relatively high similarity to Ftr1 from yeast (18.07% identical amino acid residues, 13.86% strongly similar amino acid residues, and 9.90% weakly similar amino acid residues). Interestingly, the entire *S. elongatus* PCC 7942 genome does not contain FET3 homologous genes/proteins. This can mean that either Ftr1 interacts with another yet unidentified ferrooxidase protein or that Ftr1 is part of the Irp transport system, which may be energised by IrpB, which can be assumed to have the capability to transfer and donate electrons, because it is a multiheme cytochrome c-type protein.

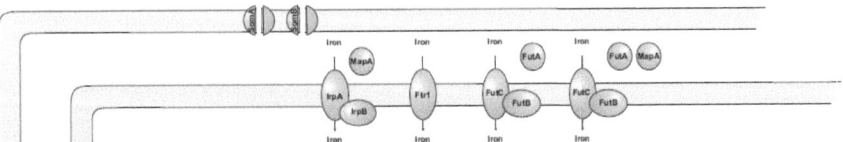

Figure 4.3: Model of the iron acquisition apparatus of *S. elongatus* PCC 7942 under conditions of iron starvation. Abbreviations are given in the previous legends. The arrangement of the iron assimilation systems in the cytoplasmic membrane represents different possible arrangements, which require further experimental support. In total, *S. elongatus* PCC 7942 probably has three different iron uptake systems: the Fut system, the Irp system, and the Ftr1 system. Whereas the Fut system is the house-keeping system, transcripts encoding subunits of the other two systems become expressed at elevated amounts under conditions of iron starvation. Since MapA in part resembles extracellular solute-binding proteins, it may very well function as such for one of the identified iron transport systems.

Related to acquisition of extracellular iron compounds, it is tempting to speculate that the enhanced expression of *somB(1)* and *somA(2)* as well as the reduced expression of *somB(2)* is directly related to this process. *S. elongatus* PCC 7942 has a total of four *som* genes, whose gene products are known to form porin-like β-barrel structures in the outer membrane, which might be connected to the S-layer. Since it is well-known that the *som* genes are expressed at different levels in response to different stress conditions, e.g. nitrogen starvation, the different expression of Som proteins under iron starvation may reflect different needs of permeability of the outer membrane, which directly account for a different cellular need for iron. This speculation gains further indirect support, since at least one out of two transcriptionally up-regulated *som* genes localises direct next to the *irpAB* operon on the chromosome. Thus, it can be concluded that *S. elongatus* PCC 7942 has three different iron uptake systems among which the Irp system showed the highest up-regulation under iron-limited growth conditions.

The microarray results also showed that transcripts of the Suf system of *S. elongatus* PCC 7942 were transcribed at elevated rates under iron starvation. This system is assumed to function as an auxiliary [Fe-S] assembly system besides the housekeeping Isc system and most likely facilitates the assembly and/or repair of the oxygen-labile [Fe-S] clusters under conditions of oxidative stress and under Fe limitation (Nachin et al., 2003; Wang et al., 2004; Balasubramanian et al., 2006). This is in good agreement with the fact that the iron limitation is known to cause increased oxidative stress and that [Fe-S] centres are very prone to damage by oxidative stress (Frausto da Silva and Williams 1993).

Among the differentially-transcribed genes, there were also two clusters of genes encoding proteins modifying the electron transport chain and a gene encoding the transcriptional activator IdiB. The function of these proteins has already be described in the introduction and in the results section and thus, should only very briefly been discussed here. IdiA modifies and protects the acceptor of PS II against oxidative stress (Lax et al. 2007; Michel and Pistorius 2004). IdiB is the transcription factor for IdiA (Michel et al. 2001), and IdiC is a ferredoxin-type protein investigated in this work (Pietsch et al. 2007) and see discussion below). IsiA has been shown to form a membrane-integral Chl *a*-binding antenna around trimeric PS I (Bibby et al. 2001a; Boekema et al. 2001), and *isiB* corresponds to the cyanobacterial flavodoxin (Burnap et al. 1993) replacing in part ferredoxin under conditions of iron starvation (Smillie 1965). The function of IsiC, which resembles hydrolase-type enzymes, is still unknown.

Another interesting finding of the DNA microarray results was related to the mechanisms of cellular energisation. As pointed out before, iron starvation inevitably leads to a progressive decay of metabolic functions and thus, to a decrease of reducing equivalents as well as of energy equivalents. In this situation, it is interesting to see that *S. elongatus* PCC 7942 taps new resources by the enhanced expression of two proteins, the acetate kinase AckA and the phosphoglycerate mutase Pgam (Figure 4.4), which are involved in acetate metabolism. So far, the acetate metabolism of cyanobacteria has not gained much attention, because the fermentative metabolism including the acetate metabolism has been considered to mainly serve maintenance purposes rather than growth (van der Oost et al. 1989). However, in iron-starved *S. elongatus* PCC 7942 the enhanced expression of Pgam and AckA puts strong metabolic emphasize on the acetate metabolism. This way it might strengthen a novel source for substrate level phosphorylation by the synthesis of ATP from acetyl phosphate. Whereas the activity of Pgam has the potential to enhance the withdrawal of reduced C3 units from the Calvin cycle into glycolysis, an enhanced activity of the acetatekinase AckA accepts acetyl phosphate, converts it to acetate with subsequent synthesis of ATP. Whether the enhanced expression of an N-acetyltransferase gene is related to a higher requirement for the conversion of acetyl-CoA from glycolysis to acetyl phosphate, remains to be investigated. By increasing the activity of the acetate metabolism, S. elongatus is able to mobilise its endogenous resources for ATP and NADPH+H$^+$ synthesis without being dependent on a functional respiratory electron transport. Moreover, due to this modification S. elongatus PCC 7942 might be able to minimize imbalances in the NADPH+H$^+$/ATP ratio and/or the C/N ratio under iron limitation.

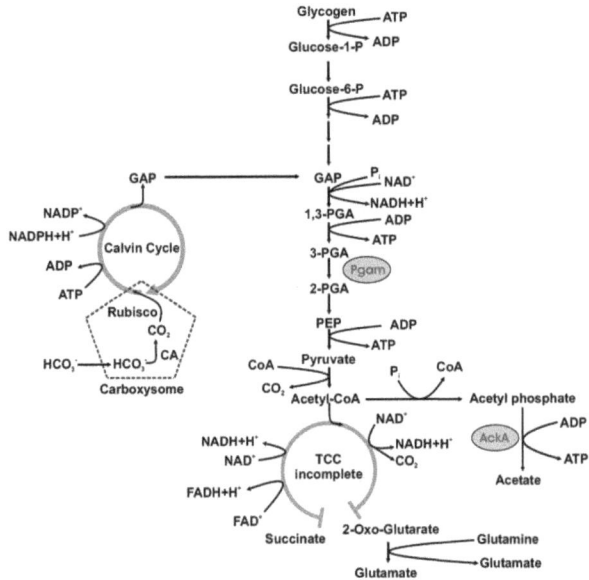

Figure 4.4: Acetate metabolism of S. elongatus PCC 7942. Under conditions of iron starvation the transcription of two genes, *pgam* and *ackA* encoding major enzymes of the acetate metabolism was found to be strongly increased (Nodop et al. 2008).

In addition to these six gene clusters, a number of other transcripts are either up- or down-regulated under iron starvation. In general, it can be stated that transcripts encoding proteins of the phycobilisomes as well as PS II and PS I-related proteins were down-regulated. In contrast, transcripts encoding subunit IV of the NDH-1 complex and various cytochrome oxidase subunits were found at elevated concentrations. Other major changes were seen for transcripts encoding major proteins of C- and N-metabolism. The same is true for transcripts for some typical stress proteins, e.g. a down-regulation of the iron superoxide dismutase and the heme-containing catalase-peroxidase KatG. In contrast, an up-regulation of the high light-induced protein C HliC and the NblA protein, which is involved in degradation of phycobilisomes, was observed.

When these transcript changes were compared between WT and the IdiB-free mutant, it became obvious that IdiB regulates the expression of a concise regulon. Previously, it has been proven that this HTH-type transcriptional activator, which is located in an operon together with the ferredoxin-like protein IdiC lying immediately upstream of *idiA* on the chromosome (Michel et al. 1999), regulates the expression of the *idiA* mRNA (Michel et al. 2001). The genome-wide DNA microarray analysis provided evidence that IdiB is a regulator of *idiA, irpA, irpB, ackA, pgam, somA(2), somB(1)*, and *ftr1* gene expression (Nodop et al. 2008). This suggests that IdiB has a major function in the signal transduction leading to the acclimation of *S. elongatus* PCC 7942 to iron deficiency. Thus, the absence of IdiB prevents the up-regulation of the Irp and the Ftr1 system suggesting that the mutant suffers more easily from iron starvation than the WT. Furthermore, the mutant cannot protect its PS II against oxidative stress as efficiently as WT due to the absence of IdiA, and the mutant cannot adjust its C-metabolism to an enhanced activity of the acetate metabolism.

Comparing the transcriptomes of iron-starved WT and of the iron-starved *idiC*-merodiploid mutant, which has low amounts of IdiA, IdiB, and IdiC, it indeed became clear that transcripts of IdiB-regulated genes were only present in low amounts. Thus, the transcriptome of the *idiC*-merodiploid mutant MuD resembled that of the IdiB-free mutant K10. In addition, the transcript alterations related to photosynthesis are slightly more pronounced in the *idiC*-merodiploid mutant than in the IdiB-free mutant and WT.

In summary, the physiological acclimation of *S. elongatus* PCC 7942 to iron starvation has been unravelled to be a measure, which predominantly aims at an improved acquisition of iron, a higher permeability of the outer membrane for extracellular solutes, a modification of the photosynthetic and respiratory electron transport chain, a potential switch to an alternative way of energy recruitment, and a controlled down-regulation of other important cellular functions such as a nitrogen assimilation.

4.2 The functional role of IdiC in the modification of electron transport under iron starvation

In the second part of my thesis the functional role of IdiC in the acclimation of *S. elongatus* PCC 7942 to iron starvation has been investigated in greater detail. Bioinformatic analyses provided evidence that IdiC has similarity to the family of thioredoxin-like ferredoxin proteins. IdiC has the highest similarity to the [2Fe-2S]-containing NuoE protein, which is the substrate-binding protein of the *E. coli* NDH-1 complex (Yagi 1998). In this respect, it has to be mentioned that so far none of the substrate-binding proteins have been identified in cyanobacteria.

In cyanobacteria the NDH-1 complex has a function in respiration (Vermaas 2001) and in at least one of the photosynthetic cyclic electron transport routes (Matthijs et al. 2002; Yeremenko et al. 2005). The results presented in my work showed that IdiC is expressed under iron starvation and also during the late growth phase. Both growth conditions are known to cause a decrease of photosynthetic linear electron transport activity and an increase in photosynthetic cyclic and respiratory electron transport. Since IdiC is expressed under these conditions and has similarity to NuoE, this hints to a function of IdiC in either photosynthetic cyclic or respiratory electron transport or maybe even in both.

As NuoE has been shown to be a [2Fe-2S]-containing protein, it was investigated whether IdiC also contains an iron cofactor. For this reason, IdiC was heterologously expressed in

E. coli and has been affinity-purified. The absorbance spectrum of the isolated protein and an ICP-OES measurement of metal ion concentrations provided evidence that IdiC contains iron. To obtain information on the structure of the cofactor, low-temperature EPR measurements were performed. Although the EPR measurements gave a signal in the g=2 region which is supportive of an [Fe-S] centre and although the shape of the signal and the multiline signal in the g=3.5-4.3 region resemble in part other [Fe-S] centres, the obtained spectrum did not allow for a clear characterisation and classification of the iron cofactor.

As already mentioned above, IdiC is expressed in the late growth phase when cells were grown in iron-sufficient medium. Therefore, the question arose whether this finding is related to an already limited supply with iron during the later growth stages. To answer this question, WT cells were grown under iron-sufficient conditions for 72 h and for another 72 h after the addition of an extra 30 µM FeCitrate. Surprisingly, immunoblot analysis showed that this treatment led to a further increase of IdiC expression as compared to the control without the addition of iron. These results imply that during the late growth the extent of IdiC expression strongly depends on the availability of iron. The high demand of IdiC expression for iron also became obvious, when IdiC was homologously expressed in *S. elongatus* PCC 7942. Following the induction of IdiC expression with IPTG in *S. elongatus* PCC 7942, the cell cultures suffered from a severe iron starvation phenotype as indicated by a bleached phenotype leading to growth arrest. The addition of an extra amount of FeCitrate in part compensated for this stress phenotype.

The evaluation of these results makes IdiC a very special protein, because this iron-containing protein is expressed under iron limitation. Moreover, it is a remarkable protein, because the supplementation of the medium with additional iron led to a further increase of IdiC expression in the late growth phase. All these results support a function of IdiC in respiration and/or photosynthetic cyclic electron transport in *S. elongatus* PCC 7942. The most unusual observation, however, is that at least two different signal transduction pathways must account for the regulation of IdiC. This conclusion is based on the observation that either iron limitation leads to an increase in expression or that expression is enhanced in the late growth phase under iron-sufficient conditions. The connecting link between both expression scenarios is that respiration is up-regulated under both growth conditions. In any way, the superior signal must arise from the fact that cells become carbohydrate-rich and rich in other endogenous substrates, and that respiration becomes a major source for ATP synthesis relative to ATP formation by photosynthesis.

This important function of IdiC in respiration is further emphasised by the fact that it has been impossible to generate a fully-insertionally inactivated IdiC-free mutant implying that IdiC has to be classified an essential protein for the viability of *S. elongatus* PCC 7942, even under regular growth conditions. Due to my results, it may become even more important under nutrient limitation and in the late growth phase.

Since all the results point to a function of IdiC in respiration and/or photosynthetic cyclic electron transport around PS I, measurements of photosynthetic and respiratory activities have been performed. As expected, photosynthetic activities were found to be lower, especially under iron-limiting conditions in the mutant as compared to WT. This is most likely due to the fact that iron starvation favors respiration as well as photosynthetic cyclic electron over photosynthetic linear electron transport from water to NADP⁺ transport and both, respiration as well as photosynthetic cyclic electron transport, require the NDH-1 complex. In addition, the lower photosynthetic activity of MuD is based on the inability of the mutant to improve iron acquisition via enhanced expression of the Irp and the Ftr1 system, and the fact that the IdiC mutant contains lower amounts of IdiA protecting PS II. Measurements of oxygen consumption with intact cells in the dark indicated that IdiC has a function in respiration as shown in chapter 3.5.3, since the oxygen consumption based on endogenous substrate decomposition was lower in the *idiC*-merodiploid mutant than in WT.

Finally, it can be stated that although the components of the photosynthetic and respiratory electron transport are well characterised, many open questions still exist on how both activities are coordinated and which additional proteins are involved in the transition of the different forms of electron transport. From the results presented in my thesis, it can be assumed that IdiC plays an important role in the transition from a predominantly photosynthetic electron transport to an enhanced respiratory and photosynthetic cyclic electron transport around PS I as being the case in the late growth phase or under specific nutrient-limiting growth conditions. Thus, the NDH-1-dependent cyclic electron transport pathway with the suggested participation of IdiC might be the major one. A model of the suggested function of IdiC in respiratory and photosynthetic cyclic electron transport is given in Figure 4.5.

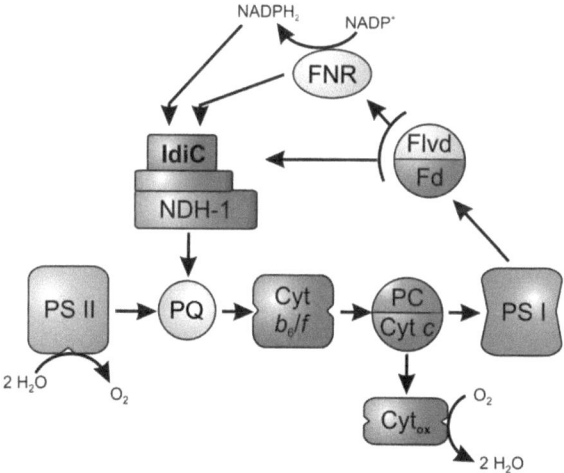

Figure 4.5: Model of the possible function of IdiC in respiratory electron transport and in photosynthetic cyclic electron transport around PS I in *S. elongatus* PCC 7942. It is suggested that IdiC, which has similarity to NuoE of the peripheral part of the NDH-1 complex from *E. coli*, is a component of the NDH-1 complex in *S. elongatus* PCC 7942. IdiC might have a function in respiration and also in NDH-1-dependent cyclic electron transport around PS I. A direct interaction of NADPH+H⁺ with the NDH-1 complex with participation of IdiC could be imagined and thus, mediating electron flow from NADPH+H⁺ to O₂ via one of the cytochrome oxidases (Cyt$_{ox}$). In photosynthetic cyclic electron transport around PS I electrons from PS I are transferred to Fd/flavodoxin (Fldv) and are suggested to be directly transported to IdiC or via FNR to IdiC.

5 Summary

The absolute requirement for iron as cofactor of oxygenic photosynthetic life-style is in sharp contrast to the severely limited bioavailability of iron. This situation is in part caused by the fact that the photosynthetic and the respiratory electron transport system contain a fairly high number of iron-containing compounds.

Transcript-profiling of *S. elongatus* PCC 7942 wild type (WT) grown under iron-deficient and iron-sufficient conditions revealed that the acclimation to iron starvation includes the enhanced expression of genes lying within six regions of the cyanobacterial chromosomes.

1. The *irp* region consisting of the genes *irpA*, *irpB*, and *somB(1)*.
2. The *fut* gene region consisting of the genes *futA*, *futB*, and *mapA*.
3. The *suf* gene region consisting of the genes *sufB*, *sufC*, *sufD*, *sufR*, and *sufS*.
4. The *isi* gene region consisting of the genes *isiA*, *isiB*, and *isiC*.
5. The *idi* gene region consisting of the genes *idiA*, *idiB*, and *idiC*.
6. The acetate metabolism region including the genes *ackA* and *pgam*.

The enhanced and differentially-regulated expression of many other genes leads to a concerted acclimatory response, which can be summarised as follows. *S. elongatus* PCC 7942 strongly improves its overall capacity of iron uptake by an increased expression of the house-keeping Fut system and two further novel putative iron uptake systems, IrpAB and Ftr1. Moreover, the permeability of the outer membrane becomes up-shifted by a modified expression of outer membrane porins. The sophisticated modification of the electron transport system by the expression of members of the Idi and Isi protein families leading to a decreased photosynthetic linear and an increased photosynthetic cyclic and respiratory electron transport as well as the potential switch to an alternative recruitment of ATP by enhancing the activity of the acetate metabolism contributes to the acclimatory process to iron limitation. This process is complemented by the controlled up-regulation of aerobic respiration via cytochrome oxidases and NDH-1 subunits and a down-regulation of other important cellular functions such as the assimilation of nitrogen.

Considering the complex regulatory cascades and the concomitant regulatory processes, which are necessary to coordinate the acclimation to iron starvation, it has been an important result that the transcript profiling of an IdiB-free mutant, IdiB being the transcriptional regulator of IdiA, led to the identification of a concise IdiB regulon. This operon comprises the genes *idiA*, *irpA*, *irpB*, *ackA*, *pgam*, *somA(2)*, *somB(1)*, and *ftr1*.

The transcript profiling approach was completed by including an IdiC mutant, which has a low amount of IdiA, IdiB, and IdiC under iron limitation as compared to WT. Under iron-deficient growth conditions the down-regulation of photosynthesis-related transcripts was more severe as compared to the other two strains. This result suggests that IdiC plays an important role in the acclimatory process to iron starvation.

Summary

IdiC is 20.5 kDa protein, which has strong similarity to thioredoxin-like [2Fe-S] ferredoxins. The highest similarity is to NuoE, which is one of the substrate-binding subunits of the NDH-1 complex of *E. coli*. Since the substrate-binding subunits of the cyanobacterial NDH-1 complex have so far not been identified, IdiC represents a good candidate for one of those subunits. In cyanobacteria, the NDH-1 complex has been shown to have a function in respiration and in at least one of the photosynthetic cyclic electron transport routes around PS I. The biochemical characterisation of the IdiC protein showed that it is an iron protein, which is loosely attached to the thylakoid membrane and which is expressed at elevated rates under iron starvation and during the late growth phase. These two growth conditions are known to lead to an increased photosynthetic cyclic electron transport and respiratory activity. These results support a function of IdiC in the cyanobacterial NDH-1 complex. Moreover, comparative measurements of the photosynthetic and respiratory activity in WT and the *idiC*-merodiploid mutant are in agreement with such a function of IdiC.

6 Acknowledgements

First I would like to thank PD Dr. Klaus-Peter Michel for excellent supervision of this work. I also would like to make a special reference to Prof. Dr. Elfriede K. Pistorius for her great support and several productive discussions. Moreover, I thank Prof. Dr. Dorothee Staiger for her support and the opportunity to perform this work at the Chair of Molecular Cell Physiology of the Bielefeld University.

Additionally, I am very thankful to our cooperation partners. I would like to thank Prof. Dr. Walter J. Horst (Institute of Plant Nutrition, Hanover University) for performing the ICP-EOS measurements, Marta J. Kopczak (Chair of Plant Biochemistry, Bochum University) for her support in the purification of the PS I complexes, Dr. Anke Nodop (Chair of Molecular Cell Physiology, Bielefeld University) for processing the micro array analyses, and Dr. Edward Reijerse (Max Planck Institute for Bioinorganic Chemistry, Mülheim/Ruhr) for realising the EPR measurements.

7 References

Allen, J. F. (1995). "Thylakoid protein phosphorylation, state1-state 2 transitions, and photosystem stoichiometry adjustment: Redox control at multiple levels of gene expression." Physiol Plantarum 93: 196-205.

Allen, J. F., K. Alexciev, et al. (1995). "Photosynthesis. Regulation by redox signalling." Curr Biol 5(8): 869-872.

Allen, J. F. and A. Nilsson (1997). "Redox signalling and the structural basis of regulation of photosynthesis by protein phosphorylation." Physiol Plantarum 100: 863-868.

Allen, M. M. and A. J. Smith (1969). "Nitrogen chlorosis in blue-green algae." Arch Mikrobiol 69(2): 114-20.

Altschul, S. F., T. L. Madden, et al. (1997). "Gapped BLAST and PSI-BLAST: a new generation of protein database search programs." Nucleic Acids Res 25(17): 3389-3402.

Anderson, J. M., W. S. Chow, et al. (1995). " The grand design of photosynthesis: Acclimation of the photosynthetic apparatus to environmental cues." Photosynth Res 46: 129-139.

Andersson, C. R., N. F. Tsinoremas, et al. (2000). "Application of bioluminescence to the study of circadian rhythms in cyanobacteria." Methods Enzymol 305: 527-542.

Andrews, S. C., A. K. Robinson, et al. (2003). "Bacterial iron homeostasis." FEMS Microbiol Rev 27(2-3): 215-37.

Andrizhiyevskaya, E. G., T. M. Schwabe, et al. (2002). "Spectroscopic properties of PSI-IsiA supercomplexes from the cyanobacterium *Synechococcus* PCC 7942." Biochim Biophys Acta 1556(2-3): 265-272.

Angerer, A., S. Gaisser, et al. (1990). "Nucleotide sequences of the *sfuA*, *sfuB*, and *sfuC* genes of *Serratia marcescens* suggest a periplasmic-binding-protein-dependent iron transport mechanism." J Bacteriol 172(2): 572-578.

Angerer, A., S. Gaisser, et al. (1990). "Nucleotide sequences of the sfuA, sfuB, and sfuC genes of Serratia marcescens suggest a periplasmic-binding-protein-dependent iron transport mechanism." J Bacteriol 172(2): 572-8.

Anraku, Y. and R. B. Gennis (1987). "The aerobic respiratory chain of *Escherichia coli*." Trends Biochem Sci 12: 262-266.

Ansari, A. Z., J. E. Bradner, et al. (1995). "DNA-bend modulation in a repressor-to-activator switching mechanism." Nature 374(6520): 371-5.

Aono, S., H. Kurita, et al. (1992). "Purification and characterization of a 7Fe ferredoxin from a thermophilic hydrogen-oxidizing bacterium, Bacillus schlegelii." J Biochem 112(6): 792-5.

Archibald, F. (1983). "*Lactobacillus plantarum*, an organism not requiring iron." FEMS Microbiol Lett 19: 29-32.

Ardelean, I., H. C. Matthijs, et al. (2002). "Unexpected changes in photosystem I function in a cytochrome c_6-deficient mutant of the cyanobacterium *Synechocystis* PCC 6803." FEMS Microbiol Lett 213(1): 113-119.

Aro, E. M., I. Virgin, et al. (1993). "Photoinhibition of photosystem II. Inactivation, protein damage and turnover." Biochem Biophys Acta 1143: 113-134.

Asada, K. (1994). Production and action of active oxygen species in photosynthetic tissues. Causes of Photooxidative Stress and Amelioration of Defense Systems in Plants. C. H. Foyer and C. Mullineaux. Boca Raton, U.S.A., CRC Press: 77-104.

Asada, K. (1999). "Responses to active oxygens, strong and weak lights, an overview." Tanpakushitsu Kakusan Koso 44(15 Suppl): 2230-2231.

Asada, K. (2000). "The water-water cycle as alternative photon and electron sinks." Philos Trans R Soc Lond B Biol Sci 355(1402): 1419-1431.

Badger, M. R., D. Hanson, et al. (2002). "Evolution and diversity of CO_2 concentrating mechanisms in cyanobacteria." Funct Plant Biol 29: 161-173.

Badger, M. R. and G. D. Price (2003). "CO_2 concentrating mechanisms in cyanobacteria: molecular components, their diversity and evolution." J Exp Bot 54(383): 609-622.

Bagchi, S. N., T. Bitz, et al. (2007). "A *Synechococcus elongatus* PCC 7942 mutant with a higher tolerance towards the herbicide bentazone also confers resistance to sodium chloride stress." Photosynth Res 92: 87-101.

Bagchi, S. N., E. K. Pistorius, et al. (2003). "A *Synechococcus* sp. PCC 7942 mutant with a higher tolerance towards bentazone." Photosynth Res 75: 171-182.

Balasubramanian, R., G. Shen, et al. (2006). "Regulatory roles for IscA and SufA in iron homeostasis and redox stress responses in the cyanobacterium *Synechococcus* sp. strain PCC 7002." J Bacteriol 188(9): 3182-3191.

Barber, J., J. Nield, et al. (2006). Accessory Chlorophyll Proteins in Cyanobacterial Photosystem I. Photosystem I: The Light-Driven Plastocyanin:Ferredoxin Oxidoreductase. J. H. Golbeck. Dordrecht, The Netherlands, Springer: 99-117.

Battchikova, N. (2008). Proteomic approaches to study acclimation of *Synechocystis* cells to low CO_2 environment. ESF Research Conference: Molecular Bioenergetics of Cyanobacteria towards Systems Biology Level of Understanding, Barcelona, Spain.

Battchikova, N., P. Zhang, et al. (2005). "Identification of NdhL and Ssl1690 (NdhO) in NDH-1L and NDH-1M complexes of *Synechocystis* sp. PCC 6803." J Biol Chem 280(4): 2587-2595.

Behrenfeld, M. J. and Z. S. Kolber (1999). "Widespread iron limitation of phytoplankton in the south pacific ocean." Science 283(5403): 840-843.

Berger, S., U. Ellersiek, et al. (1993). "Immunopurification of a subcomplex of the NAD(P)H-plastoquinone-oxidoreductase from the cyanobacterium *Synechocystis* sp. PCC 6803." FEBS Lett 326: 249-250.

Bhaya, D., R. Schwarz, et al. (2000). Molecular responses to environmental stress. The Ecology of Cyanobacteria: Their Diversity in Time and Space. B. A. Whitton and M. Potts. Dordrecht, London & Boston, Kluwer Academic Publishers, U.S.A.: 397-442.

Bibby, T. S., J. Nield, et al. (2001a). "Iron deficiency induces the formation of an antenna ring around trimeric photosystem I in cyanobacteria." Nature 412(6848): 743-745.

Bibby, T. S., J. Nield, et al. (2001b). "Three-dimensional model and characterisation of the iron-stress induced CP43'-photosystem I supercomplex isolated from the cyanobacterium *Synechocystis* PCC 6803." J Biol Chem 22: 22-30.

Bibby, T. S., J. Nield, et al. (2001c). "Oxyphotobacteria: Antenna ring around photosystem I." Nature 413(6856): 590.

Block, M. A. and A. R. Grossman (1988). "Identification and Purification of a Derepressible Alkaline Phosphatase from Anacystis nidulans R2." Plant Physiol 86(4): 1179-1184.

Bockholt, R., B. Masepohl, et al. (1995). "Partial amino acid sequence of an L-amino acid oxidase from the cyanobacterium Synechococcus PCC6301, cloning and DNA sequence analysis of the aoxA gene." Biochim Biophys Acta 1264(3): 289-93.

Bockholt, R., B. Masepohl, et al. (1995). "Partial amino acid sequence of an L-amino acid oxidase from the cyanobacterium *Synechococcus* PCC6301, cloning and DNA sequence analysis of the *aoxA* gene." Biochim Biophys Acta 1264(3): 289-293.

Bockholt, R., G. Scholten-Beck, et al. (1996). "Construction and partial characterization of an L-amino acid oxidase- free *Synechococcus* PCC 7942 mutant and localization of the L-amino acid oxidase in the corresponding wild type." Biochim Biophys Acta 1307(1): 111-121.

Boekema, E. J., A. Hifney, et al. (2001). "A giant chlorophyll-protein complex induced by iron deficiency in cyanobacteria." Nature 412(6848): 745-748.

Boyer, G. L., A. H. Gillam, et al. (1987). Iron chelation and uptake. The Cyanobacteria. P. Fay. Amsterdam, New York, Oxford, Elsevier: 415-436.

Bradford, M. M. (1976). "A rapid and sensitive method for the quantitation of microgram quantities of protein utilizing the principle of protein-dye binding." Anal Biochem 72: 248-254.

Braun, V. (1998). "Protein structure: Pumping iron through cell membranes." Science 282(5397): 2202-2203.

Braun, V. (2001). "Iron uptake mechanisms and their regulation in pathogenic bacteria." Int J Med Microbiol 291(2): 67-79.

Braun, V., R. Gross, et al. (1983). "Plasmid and chromosomal mutants in the iron(III)-aerobactin transport system of Escherichia coli. Use of streptonigrin for selection." Mol Gen Genet 192(1-2): 131-9.

Bricker, T. M. and L. K. Frankel (2002). "The structure and function of CP47 and CP43 in photosystem II." Photosynth Res 72(2): 131-146.

Brown, N. L., J. V. Stoyanov, et al. (2003). "The MerR family of transcriptional regulators." FEMS Microbiol Rev 27(2-3): 145-63.

Brown, N. L., J. V. Stoyanov, et al. (2003). "The MerR family of transcriptional regulators." FEMS Microbiol Rev 27(2-3): 145-163.

Bruce, D. and J. Biggins (1985). "Mechanism of the light-state transition in photosynthesis. V. 77 K linear dichroism of *Anacystis nidulans* in state 1 and state 2." Biochem Biophys Acta 810: 295-310.

Bruce, D., J. Biggins, et al. (1985). "Mechanism of the light-state transition in photosynthesis. IV. Picosecond fluorescence spectroscopy of *Anacystis nidulans* and *Porphiridium cruentum* in state 1 and state 2 at 77 K." Biochem Biophys Acta 806: 237-246.

Brune, I., A. Becker, et al. (2006). "Under the influence of the active deodorant ingredient 4-hydroxy-3-methoxybenzyl alcohol, the skin bacterium *Corynebacterium jeikeium* moderately responds with differential gene expression." J Biotechnol 127(1): 21-33.

Burnap, R. L., T. Troyan, et al. (1993). "The highly abundant chlorophyll-protein complex of iron-deficient *Synechococcus* sp. PCC 7942 (CP43') is encoded by the *isiA* gene." Plant Physiol 103(3): 893-902.

Carr, N. G. and B. A. Whitton (1982). The Biology of Cyanobacteria. Oxford, UK, Blackwell Scientific Publishers.

Carter, K. R., J. Rawlings, et al. (1980). "Purification and characterization of a ferredoxin from Rhizobium japonicum bacteroids." J Biol Chem 255(9): 4213-23.

Chen, C. Y., S. A. Berish, et al. (1993). "The ferric iron-binding protein of pathogenic *Neisseria* spp. functions as a periplasmic transport protein in iron acquisition from human transferrin." Mol Microbiol 10(2): 311-318.

Chitnis, V. P. and P. R. Chitnis (1993). "PsaL subunit is required for the formation of photosystem I trimers in the cyanobacterium *Synechocystis* sp. PCC 6803." FEBS Lett 336(2): 330-334.

Cody, G. D., N. Z. Boctor, et al. (2000). "Primordial carbonylated iron-sulfur compounds and the synthesis of pyruvate." Science 289(5483): 1337-1340.

Collier, J. L. and A. R. Grossman (1994). "A small polypeptide triggers complete degradation of light-harvesting phycobiliproteins in nutrient-deprived cyanobacteria." Embo J 13(5): 1039-47.

Cosgrove, K., G. Coutts, et al. (2007). "Catalase (KatA) and alkyl hydroperoxide reductase (AhpC) have compensatory roles in peroxide stress resistance and are required for survival, persistence, and nasal colonization in *Staphylococcus aureus*." J Bacteriol 189(3): 1025-1035.

De Las Rivas, J., M. Balsera, et al. (2004). "Evolution of oxygenic photosynthesis: genome-wide analysis of the OEC extrinsic proteins." Trends Plant Sci 9(1): 18-25.

Deng, Y., J. Ye, et al. (2003). "Effects of low CO_2 on NAD(P)H dehydrogenase, a mediator of cyclic electron transport around photosystem I in the cyanobacterium *Synechocystis* PCC6803." Plant Cell Physiol 44(5): 534-540.

DeRisi, J. L., V. R. Iyer, et al. (1997). "Exploring the metabolic and genetic control of gene expression on a genomic scale." Science 278(5338): 680-686.

DeRuyter, Y. S. and P. Fromme (2008). Molecular structure od the photosynthetic apperatus. The cyanobacteria. F. E. and A. Herrero, Caister Academic Press.

Dietz, K. J., T. Stork, et al. (2002). The Role of Peroxiredoxins in Oxygenic Photosynthesis of Cyanobacteria and Higher Plants: Peroxide Detoxification or Redox Sensing? Photoprotection, Photoinhibition, Gene Regulation, and Environment. B. Demming-Adams, W. W. Adams III and A. K. Mattoo. Dordrecht, The Netherlands, Springer: 303-319.

Dondrup, M., A. Goesmann, et al. (2003). "EMMA: a platform for consistent storage and efficient analysis of microarray data." J Biotechnol 106(2-3): 135-146.

Drechsel, H. and G. Winkelmann (1997). Iron chelation and siderophores. Transition Metals in Microbial Metabolism. G. Winkelmann and C. J. Carrano. Amsterdam, Harwood Academic Publishers: 1-49.

Dudoit, S., Y. Yang, et al. (2002). "Statistical methods for identifying differentially expressed genes in replicated cDNA microarray experiments." Stat Sin 12: 111-139.

Duhring, U., I. M. Axmann, et al. (2006). "An internal antisense RNA regulates expression of the photosynthesis gene *isiA*." Proc Natl Acad Sci U. S. A. 103(18): 7054-7058.

Durham, K. A. and G. S. Bullerjahn (2002). "Immunocytochemical localization of the stress-induced DpsA protein in the cyanobacterium *Synechococcus* sp. strain PCC 7942." J Basic Microbiol 42(6): 367-372.

Durham, K. A., D. Porta, et al. (2003). "Expression of the iron-responsive *irpA* gene from the cyanobacterium *Synechococcus* sp. strain PCC 7942." Arch Microbiol 179(2): 131-114.

Dwivedi, K., A. Sen, et al. (1997). "Expression and mutagenesis of the *dpsA* gene of *Synechococcus* sp. PCC 7942, encoding a DNA-binding protein involved in oxidative stress protection." FEMS Microbiol Lett 155: 85-91.

Eisenhut, M., E. A. von Wobeser, et al. (2007). "Long-term response toward inorganic carbon limitation in wild type and glycolate turnover mutants of the cyanobacterium *Synechocystis* sp. strain PCC 6803." Plant Physiol 144(4): 1946-1959.

El-Mohsnawy, E. (2007). "Evaluation of monomeric and trimeric PS1 in native systems and in semiartificial systems for biohydrogen production."

Elliott, J. I. and L. G. Ljungdahl (1982). "Isolation and characterization of an Fe,-S8 ferredoxin (ferredoxin II) from Clostridium thermoaceticum." J Bacteriol 151(1): 328-33.

Elstner, E. F. (1990). Der Sauerstoff. Biochemie, Biologie, Medizin. Heidelberg, Germany, Spektrum Akademischer Verlag.

Engels, A., U. Kahmann, et al. (1997). "Isolation, partial characterization and localization of a dihydrolipoamide dehydrogenase from the cyanobacterium *Synechocystis* PCC 6803." Biochim Biophys Acta 1340(1): 33-44.

Engels, A. and E. K. Pistorius (1997). "Characterization of a gene encoding dihydrolipoamide dehydrogenase of the cyanobacterium *Synechocystis* sp. strain PCC 6803." Microbiology 143 (Pt 11): 3543-3553.

Engels, D. H., A. Engels, et al. (1992). "Isolation and partial characterization of a L-amino acid oxidase and of photosystem II complexes from the cyanobacterium *Synechococcus* PCC 7942." Bioscience C 47: 859-866.

Ernst, J. F., R. L. Bennett, et al. (1978). "Constitutive expression of the iron-enterochelin and ferrichrome uptake systems in a mutant strain of *Salmonella typhimurium*." J Bacteriol 135(3): 928-934.

Escolar, L., J. Perez-Martin, et al. (1999). "Opening the iron box: transcriptional metalloregulation by the Fur protein." J Bacteriol 181(20): 6223-6229.

Exss-Sonne, P. (2000). Untersuchungen zur Lokalisation und Funktion von IdiA in dem thermophilen Cyanobakterium *Synechococcus elongatus* und dem mesophilen Cyanobakterium *Synechococcus* PCC 7942. Fakultät für Biologie. Bielefeld, Bielefeld: 133.

Exss-Sonne, P., J. Tölle, et al. (2000). "The IdiA protein of *Synechococcus* sp. PCC 7942 functions in protecting photosystem II under oxidative stress." Photosynth Res 63: 145-157.

Falk, S., G. Samson, et al. (1995). "Functional analysis of the iron-stress induced CP43' polypeptide of PS II in the cyanobacterium *Synechococcus* sp. PCC 7942." Photosynth Res 45: 51-60.

Fay, P. (1983). The Blue-Greens. Baltimoore, U.S.A., Edward Arnold.

Fecht-Christoffers, M. M., H. Führs, et al. (2006). "The Role of Hydrogen Peroxide-Producing and Hydrogen Peroxide-Consuming Peroxidases in the Leaf Apoplast of Cowpea in Manganese Tolerance." Plant Physiology 140: 1451-1463.

Fellay, R., J. Frey, et al. (1987). "Interposon mutagenesis of soil and water bacteria: a family of DNA fragments designed for in vitro insertional mutagenesis of gram- negative bacteria." Gene 52(2-3): 147-154.

Ferreira, F. and N. A. Straus (1994). "Iron deprivation in cyanobacteria." J Appl Phycol 6: 199-210.

Ferreiros, C., M. T. Criado, et al. (1999). "The neisserial 37 kDa ferric binding protein (FbpA)." Comp Biochem Physiol B Biochem Mol Biol 123(1): 1-7.

Fiore, M. F., D. H. Moon, et al. (2000). "Miniprep DNA isolation from unicellular and filamentous cyanobacteria." J Microbiol Methods 39(2): 159-69.

Flores, E., A. Herrero, et al. (1982). "Production of ammonium dependent on basic L-amino acids by *Anacystis nidulans*." Arch Microbiol 131: 91-94.

Forchhammer, K. and N. Tandeau de Marsac (1994). "The PII protein in the cyanobacterium Synechococcus sp. strain PCC 7942 is modified by serine phosphorylation and signals the cellular N-status." J Bacteriol 176(1): 84-91.

Frasch, W. D. (1994). The F-type ATPase in cyanobacteria: Pivotal point in the evolution of a universal enzyme. The Molecular Biology of Cyanobacteria. D. A. Bryant. Dordrecht, Boston & London, Kluwer Academic Publishers. 4: 361-380.

Frausto da Silva, J. J. R. and R. J. P. Williams (1993). The Biological Chemistry of the Elements. The Inorganic Chemistry of Life. Oxford, Clarendon Press.

Frias, J. E., A. Merida, et al. (1993). "General distribution of the nitrogen control gene *ntcA* in cyanobacteria." J Bacteriol 175(17): 5710-5713.

Friedrich, T. (1998). "The NADH:ubiquinone oxidoreductase (complex I) from *Escherichia coli*." Biochim Biophys Acta 1364(2): 134-146.

Friedrich, T. and D. Scheide (2000). "The respiratory complex I of bacteria, archaea and eukarya and its module common with membrane-bound multisubunit hydrogenases." FEBS Lett 479(1-2): 1-5.

Friedrich, T., K. Steinmüller, et al. (1995). "The proton-pumping respiratory complex I of bacteria and mitochondria and its homologue in chloroplasts." FEBS Lett 367: 107-111.

Friedrich, T. and H. Weiss (1997). "Modular evolution of the respiratory NADH:ubiquinone oxidoreductase and the origin of its modules." J Theor Biol 187(4): 529-40.

Fromme, P. and H. T. Witt (1998). "Improved isolation and crystallization of photosystem I for structural analysis." Biochem Biophys Acta 1365: 175-184.

Fry, I. V., M. Huflejt, et al. (1986). "The role of respiration during adaptation of the freshwater cyanobacterium *Synechococcus* 6311 to salinity." Arch Biochem Biophys 244(2): 686-691.

Fuangthong, M., A. F. Herbig, et al. (2002). "Regulation of the *Bacillus subtilis fur* and *perR* genes by PerR: not all members of the PerR regulon are peroxide inducible." J Bacteriol 184(12): 3276-3286.

Fulda, S., F. Huang, et al. (2000). "Proteomics of *Synechocystis* sp. strain PCC 6803: Identification of periplasmic proteins in cells grown at low and high salt concentrations." Eur J Biochem 267(19): 5900-5907.

Fulda, S., S. Mikkat, et al. (2006). "Proteome analysis of salt stress response in the cyanobacterium *Synechocystis* sp. strain PCC 6803." Proteomics 6(9): 2733-2745.

Fulda, S., S. Mikkat, et al. (1999). "Isolation of salt-induced periplasmic proteins from *Synechocystis* sp. strain PCC 6803." Arch Microbiol 171(3): 214-217.

Gantt, E. (1994). Supramolecular membrane organisation. The Molecular Biology of Cyanobacteria. D. A. Bryant. Dordrecht, Boston, London, The Netherlands, Kluwer Academic Publishers. **4:** 119-138.

Gau, A. E., A. Heindl, et al. (2007). "L-amino acid oxidases with specificity for basic amino acids in cyanobacteria." Z Naturforsch [C](62): 273-284.

Geider, R. J. and J. La Roche (1994). "The role of iron in phytoplankton photosynthesis and the potential for iron limitation of primary productivity in the sea." Photosynth Res 39: 275-301.

Geiss, U., J. Vinnemeier, et al. (2001). "Detection of the *isiA* gene across cyanobacterial strains: potential for probing iron deficiency." Appl Environ Microbiol 67(11): 5247-5253.

Ghassemian, M. and N. A. Straus (1996). "Fur regulates the expression of iron-stress genes in the cyanobacterium *Synechococcus* sp. strain PCC 7942." Microbiology 142(Pt 6): 1469-1476.

Gibney, B. R., S. E. Mulholland, et al. (1996). "Ferredoxin and ferredoxin-heme maquettes." Proc Natl Acad Sci U S A 93(26): 15041-6.

Golbeck, J. H. (1994). Photosystem I in cyanobacteria. The Molecular Biology of Cyanobacteria. D. A. Bryant. Dordrecht, Boston & London, Kluwer Academic Publishers. **4:** 320-360.

Golding, A. J. and G. N. Johnson (2003). "Down-regulation of linear and activation of cyclic electron transport during drought." Planta 218(1): 107-114.

Golitsyn, V. M., V. L. Tetenkin, et al. (1995). "Spectral properties of cyanobacterium *Synechocystis* sp. PCC 6803 mutants lacking photosystem II activity." Biochemistry (Moscow) 60: 359-362.

Gomez-Lojero, C., B. Perez-Gomez, et al. (2003). "Interaction of ferredoxin:NADP+ oxidoreductase with phycobilisomes and phycobilisome substructures of the cyanobacterium Synechococcus sp. strain PCC 7002." Biochemistry 42(47): 13800-11.

Grigorieff, N. (1999). "Structure of the respiratory NADH:ubiquinone oxidoreductase (complex I)." Curr Opin Struct Biol 9(4): 476-483.

Grimme, L. H. and N. K. Boardman (1972). "Photochemical activation of a particle fraction P1 obtained from the green alga *Chlorella fusca*." Biochem Biophys Res Commun 49: 1617-1623.

Grossman, A. R., M. R. Schaefer, et al. (1994). The responses of cyanobacteria to environmental conditions: Light and nutrients. The Molecular Biology of Cyanobacteria. D. A. Bryant. Dordrecht, Boston & London, Kluwer Academic Publishers. **4:** 641-675.

Guedeney, G., S. Corneille, et al. (1996). "Evidence for an association of ndh B, ndh J gene products and ferredoxin-NADP-reductase as components of a chloroplastic NAD(P)H dehydrogenase complex." FEBS Lett 378(3): 277-80.

Guerinot, M. L. and Y. Yi (1994). "Iron: Nutritious, noxious, and not readily available." Plant Physiol 104(3): 815-820.

Guikema, J. A. and L. A. Sherman (1984). "Influence of iron deprivation on the membrane composition of *Anacystis nidulans*." Plant Physiol 74: 90-95.

Hagemann, M., R. Jeanjean, et al. (1999). "Flavodoxin accumulation contributes to enhanced cyclic electron flow around photosystem I in salt-stressed cells of *Synechocystis* sp. strain PCC 6803." Physiol Plant 105: 670-678.

Hanahan, D. and M. Meselson (1983). "Plasmid screening at high colony density." Methods Enzymol 100: 333-42.

Hansel, A., F. Pattus, et al. (1998). "Cloning and characterization of the genes coding for two porins in the unicellular cyanobacterium *Synechococcus* PCC 6301." Biochim Biophys Acta 1399(1): 31-39.

Hantke, K. (1981). "Regulation of ferric iron transport in *Escherichia coli* K12: isolation of a constitutive mutant." Mol Gen Genet 182(2): 288-292.

Hantke, K. and V. Braun (1998). Control of bacterial iron transport by regulatory proteins. Metal Ions in Gene Regulation. S. Silver and W. Walden. New York, U.S.A., International Thomson Publishing: 45-76.

Hart, S. E., B. G. Schlarb-Ridley, et al. (2005). "Terminal oxidases of cyanobacteria." Biochem Soc Trans 33(Pt 4): 832-835.

Havaux, M., G. Guedeney, et al. (2005). "The chlorophyll-binding protein IsiA is inducible by high light and protects the cyanobacterium *Synechocystis* PCC6803 from photooxidative stress." FEBS Lett 579(11): 2289-2293.

Heinecke, D. (2001). Photosynthesis: Dark reactions. Encyclopedia of Life Sciences. London, Nature Publishing Group: 1-4.

Helmann, J. D. (1998). Metal cation regulation in Gram-positive bacteria. Metal Ions in Gene Regulation. S. Silver and W. Walden. New York, U.S.A., International Thomson Publishing: 45-76.

Herbert, S. K., R. E. Martin, et al. (1995). "Light adaptation of cyclic electron transport through photosystem I in the cyanobacterium *Synechococcus* sp. PCC 7942." Photosynth Res 46: 277-285.

Herbert, S. K., G. Samson, et al. (1992). "Characterization of damage to photosystems I and II in a cyanobacterium lacking detectable iron superoxide dismutase activity." Proc Natl Acad Sci U S A 89(18): 8716-8720.

Herranen, M., N. Battchikova, et al. (2004). "Towards functional proteomics of membrane protein complexes in *Synechocystis* sp. PCC 6803." Plant Physiol 134(1): 470-481.

Herrero, A., A. M. Muro-Pastor, et al. (2001). "Nitrogen control in cyanobacteria." J Bacteriol 183(2): 411-425.

Holm, L., C. Sander, et al. (1994). "LexA repressor and iron uptake regulator from *Escherichia coli*: new members of the CAP-like DNA binding domain superfamily." Protein Eng 7(12): 1449-1453.

Howitt, C. A., P. K. Udall, et al. (1999). "Type 2 NADH dehydrogenases in the cyanobacterium *Synechocystis* sp. strain PCC 6803 are involved in regulation rather than respiration." J. Bacteriol. 181(13): 3994-4003.

Huang, J. J., N. H. Kolodny, et al. (2002). "The acid stress response of the cyanobacterium *Synechocystis* sp. strain PCC 6308." Arch Microbiol 177(6): 486-493.

Hugo, N., J. Armengaud, et al. (1998). "A novel -2Fe-2S- ferredoxin from Pseudomonas putida mt2 promotes the reductive reactivation of catechol 2,3-dioxygenase." J Biol Chem 273(16): 9622-9.

Ihalainen, J. A., S. D'Haene, et al. (2005). "Aggregates of the chlorophyll-binding protein IsiA (CP43') dissipate energy in cyanobacteria." Biochemistry 44(32): 10846-10853.

Ivanov, A. G., M. Krol, et al. (2006). "Iron deficiency in cyanobacteria causes monomerization of photosystem I trimers and reduces the capacity for state transitions and the effective absorption cross section of photosystem I in vivo." Plant Physiol 141(4): 1436-1445.

References

Ivanov, A. G., Y. I. Park, et al. (2000). "Iron stress restricts photosynthetic intersystem electron transport in *Synechococcus* sp. PCC 7942." FEBS Lett 485(2-3): 173-177.

Jänsch, L. (1996). "New insights into the composition, molecular mass and stochiometry of the protein complexes of plant mitochondria." The Plant Journal 9(3): 357-368.

Jeanjean, R., J. J. van Thor, et al. (1999). Identification of plastoquinone-cytochrome b_6f reductase pathways in direct or indirect photosystem I driven cyclic electron flow in *Synechocystis* PCC 6803. The phototrophic prokaryotes. G. Peschek, W. Löffelhardt and G. Schmetterer. New York, Boston, Dordrecht, London and Moscow, Kluwer Academic/Plenum Publishers: 251-258.

Johnson, G. N. (2005). "Cyclic electron transport in C3 plants: fact or artefact?" J Exp Bot 56(411): 407-416.

Joliot, P. and A. Joliot (2006). Cyclic Electron Transfer Around Photosystem I. Photosystem I: The Light-Driven Plastocyanin:Ferredoxin Oxidoreductase. J. H. Golbeck. Dordrecht, The Netherlands, Springer. 24: 639-656.

Jordan, P., P. Fromme, et al. (2001). "Three-dimensional structure of cyanobacterial photosystem I at 2.5 A resolution." Nature 411(6840): 909-917.

Joset, F., R. Jeanjean, et al. (1996). "Dynamics of the response of cyanobacteria to salt stress: Deciphering the molecular events." Physiol Plant 96: 738-744.

Jurgens, U. and J. Weckesser (1985). "The fine structure and chemical composition of the cell wall and sheath layers of cyanobacteria." Ann Inst Pasteur Microbiol 136A(1): 41-44.

Kaim, W. and B. Schwederski (1995). "Bioanorganisch Chemie."

Kammler, M., C. Schon, et al. (1993). "Characterization of the ferrous iron uptake system of Escherichia coli." J Bacteriol 175(19): 6212-9.

Kaneko, T., S. Sato, et al. (1996). "Sequence analysis of the genome of the unicellular cyanobacterium *Synechocystis* sp. strain PCC 6803. II. Sequence determination of the entire genome and assignment of potential protein-coding regions (supplement)." DNA Res 3(3): 185-209.

Kaplan, A. and L. Reinhold (1999). "CO_2 concentrating mechanisms in photosynthetic microorganisms." Annu Rev Plant Physiol Plant Mol Biol 50: 539-570.

Karplus, P. A. and H. R. Faber (2004). "Structural Aspects of Plant Ferredoxin : NADP(+) Oxidoreductases." Photosynth Res 81(3): 303-15.

Katoh, H., N. Hagino, et al. (2001a). "Genes essential to iron transport in the cyanobacterium *Synechocystis* sp. strain PCC 6803." J Bacteriol 183(9): 2779-2784.

Katoh, H., N. Hagino, et al. (2001b). "Iron-binding of FutA1 subunit of an ABC-type iron transporter in the cyanobacterium *Synechocystis* sp. strain PCC 6803." Plant Cell Physiol 42(8): 823-827.

Ke, B. (2001). Photosynthesis: Photobiochemistry and Photobiophysics. Dordrecht, Boston & London, Kluwer Academic Publishers.

Kirby, S., F. Lainson, et al. (1998). "The *Pasteurella haemolytica* 35 kDa iron-regulated protein is an FbpA homologue." Microbiology 144(12): 3425-3436.

Klughammer, B., D. Sultemeyer, et al. (1999). "The involvement of NAD(P)H dehydrogenase subunits, NdhD3 and NdhF3, in high-affinity CO2 uptake in Synechococcus sp. PCC7002 gives evidence for multiple NDH-1 complexes with specific roles in cyanobacteria." Mol Microbiol 32(6): 1305-15.

Koropatkin, N., A. M. Randich, et al. (2007). "The structure of the iron-binding protein, FutA1, from *Synechocystis* 6803." J Biol Chem 282(37): 27468-27477.

Kösling, S. (1999). The complex I-homologous NAD(P)H-plastoquinone-oxidoreductase of *Synechocystis* sp. PCC 6803. The Phototrophic Prokaryotes. G. A. Peschek, Löffelhardt, W., Schmetterer, G: 211- 215.

Kotani, H. and S. Tabata (1998). "Lessons from sequencing of the genome of a unicellular cyanobacterium *Synechocystis* sp. PCC 6803." Annu Rev Plant Physiol Plant Mol Biol 49: 151-171.

Kouril, R., A. A. Arteni, et al. (2005). "Structure and functional role of supercomplexes of IsiA and photosystem I in cyanobacterial photosynthesis." FEBS Lett 579(15): 3253-3257.

Krauß, N. and W. Sänger (2001). Photosystem I. Encyclopedia of Life Sciences. London, Nature Publishing Group: 1-7.

Krieger-Liszkay, A. and A. W. Rutherford (1998). "Influence of herbicide binding on the redox potential of the quinone acceptor in photosystem II: relevance to photodamage and phytotoxicity." Biochemistry 37(50): 17339-17344.

Kunert, A. (2001). Etablierung einer Reportergensystems in *Synechocystis* sp. Stamm PCC 6803 und seine Anwendung bei der Analyse der Eisenmangel-induzierten P_{isiAB}-Aktivität. Fakultät für Biologie. Rostock, Germany, Universität Rostock: 98.

Kunert, A., J. Vinnemeier, et al. (2003). "Repression by Fur is not the main mechanism controlling the iron-inducible *isiAB* operon in the cyanobacterium *Synechocystis* sp. PCC 6803." FEMS Microbiol Lett in press.

Kunert, A., J. Vinnemeier, et al. (2003). "Repression by Fur is not the main mechanism controlling the iron-inducible *isiAB* operon in the cyanobacterium *Synechocystis* sp. PCC 6803." FEMS Microbiol Lett 227(2): 255-262.

Kutzki, C., B. Masepohl, et al. (1998). "The *isiB* gene encoding flavodoxin is not essential for photoautotrophic iron limited growth of the cyanobacterium *Synechocystis* sp. strain PCC 6803." FEMS Microbiol Lett 160(2): 231-235.

Laemmli, U. K. (1970). "Cleavage of structural proteins during the assembly of the head of bacteriophage T4." Nature 227(259): 680-685.

Larrondo, L. F., P. Canessa, et al. (2007). "Cloning and characterization of the genes encoding the high-affinity iron-uptake protein complex Fet3/Ftr1 in the basidiomycete *Phanerochaete chrysosporium*." Microbiology 153(Pt 6): 1772-1780.

Laudenbach, D. E., M. E. Reith, et al. (1988). "Isolation, sequence analysis, and transcriptional studies of the flavodoxin gene from *Anacystis nidulans* R2." J Bacteriol 170(1): 258-265.

Laudenbach, D. E. and N. A. Straus (1988). "Characterization of a cyanobacterial iron stress-induced gene similar to *psbC*." J Bacteriol 170(11): 5018-5026.

Laulhere, J. P., A. M. Laboure, et al. (1992). "Purification, characterization and function of bacterioferritin from the cyanobacterium *Synechocystis* P.C.C. 6803." Biochem J 281(Pt 3): 785-793.

Lax, J., E. J. Boekema, et al. (2007). "Structural response of photosystem 2 to iron deficiency: characterization of a new photosystem 2-IdiA complex from the cyanobacterium *Thermosynechococcus elongatus* BP-1." Biochim Biophys Acta 1767(6): 528-534.

Lengeler, J. W., G. Drews, et al. (1999). Biology of the Prokaryotes. Stuttgart, New York, Thieme.

Li, H., A. K. Singh, et al. (2004). "Differential gene expression in response to hydrogen peroxide and the putative PerR regulon of *Synechocystis* sp. strain PCC 6803." J Bacteriol 186(11): 3331-3345.

Li, H., A. K. Singh, et al. (2003). Differential gene expression and the regulation of oxidative stress in the cyanobacterium *Synechocystis* sp. PCC 6803. Functional Genomics in Cyanobacteria, Benediktbeuren, Germany.

Liberton, M., R. Howard Berg, et al. (2006). "Ultrastructure of the membrane systems in the unicellular cyanobacterium Synechocystis sp. strain PCC 6803." Protoplasma 227 (2-4): 129-38.

Lim, M. (2003). Analyse des Eisen-regulierten des *idiB* Promotors aus dem Cyanobakterium *Synechococcus* sp. PCC 7942. Fakultät für Biologie. Bielefeld, Germany, Universität Bielefeld.

Lund, P. A. and N. L. Brown (1989). "Regulation of transcription in Escherichia coli from the mer and merR promoters in the transposon Tn501." J Mol Biol 205(2): 343-53.

Lund, P. A., S. J. Ford, et al. (1986). "Transcriptional regulation of the mercury-resistance genes of transposon Tn501." J Gen Microbiol 132(2): 465-80.

Lundrigen, M. D., J. E. L. Arceneaux, et al. (1997). "Enhanced hydrogen peroxide sensitivity and altered stress protein expression in iron-starved *Mycobacterium smegmatis*." Biometals 10: 215-225.

Mann, N. H. (2000). Detecting the environment. The Ecology of Cyanobacteria. B. A. Whitton and M. Potts. Dordrecht, London & Boston, Kluwer Academic Publishers: 367-395.

Manna, P. and W. Vermaas (1997). "Lumenal proteins involved in respiratory electron transport in the cyanobacterium Synechocystis sp. PCC6803." Plant Mol Biol 35(4): 407-16.

Marjorette, M., O. Pena, et al. (1995). "The DpsA protein of *Synechococcus* sp. strain PCC 7942 is a DNA-binding hemoprotein." J Biol Chem 270(38): 22478-22782.

Martin, J. H. (1992). Iron as a limiting factor in oceanic productivity. Primary Productivity and Biogeochemical Cycles in the Sea. P. G. Falkowski and A. D. Woodhead. New York, U.S.A., Plenum Press: 123-127.

Martin, J. H., K. H. Coale, et al. (1994). "Testing the iron hypothesis in the ecosystem of the equatorial Pacific Ocean." Nature 371: 123-129.

Martin, J. H. and S. E. Fitzwater (1988). "Iron deficiency limits phytoplankton growth in the north-east Pacific subarctic." Nature 331: 341-343.

Matsuo, M., T. Endo, et al. (1998). "Isolation of a novel NAD(P)H-quinone oxidoreductase from the cyanobacterium Synechocystis PCC6803." Plant Cell Physiol 39(7): 751-5.

Matthijs, H. C., R. Jeanjean, et al. (2002). "Hypothesis: versatile function of ferredoxin-$NADP^+$ reductase in cyanobacteria provides regulation for transient photosystem I-driven cyclic electron transport." Funct Plant Biol 29: 201-210.

Melkozernov, A. N., T. S. Bibby, et al. (2003). "Time-resolved absorption and emission show that the CP43' antenna ring of iron-stressed *Synechocystis* sp. PCC6803 is efficiently coupled to the photosystem I reaction center core." Biochemistry 42(13): 3893-3903.

Meunier, P. C., M. S. Colon-Lopez, et al. (1997). "Temporal changes in state transitions and photosystem organization in the unicellular, diazotrophic cyanobacterium *Cyanothece* sp. ATCC 51142." Plant Physiol 115(3): 991-1000.

Mi, H. (1992). " Electron donation from cyclic and respiratory flows to the photosynthetic intersystem chains is mediated by pyridine nucleotide dehydrogenase in the cyanobacterium Synechocystis PCC 6803." Plant cell physiol. 33: 1233-1237.

Mi, H. (1994). "NAD(P)H Dehydrogenase-Dependent Cyclic Electron Flow around Photosystem I in the Cyanobacterium *Synechocystis* PCC 6803: a Study of Dark-Starved Cells and Spheroplasts." Plant cell physiol. 35: 163-173.

Mi, H. (1995). "Thylakoid Membrane-Bound, NADPH_Specific Pyridine Nucleotide Dehydrogenase Complex Mediates Cyclic Electron Transport in the Cyanobacterium *Synechocystis* sp. PCC 6803." Plant cell physiol. 36: 661-668.

Mi, H., T. Endo, et al. (1994). "NAD(P)H dehydrogenase-dependent cyclic electron flow around photosystem I in the cyanobacterium *Synechocystis* sp. PCC 6803: A study of dark-starved cells and spheroplasts." Plant Cell Physiol 35: 163-173.

Mi, H., T. Endo, et al. (1992). "Electron donation from cyclic and respiratory flows to the photosynthetic intersystem chain is mediated by the pyridine nucleotide dehydrogenase in the cyanobacterium Synechocystis PCC 6803." Plant Cell Physiol 33: 1233-1237.

Michel, K. P. (1996). Molekularbiologische Untersuchungen des für die Eisen- und Manganversorgung der Cyanobakterien *Synechococcus* PCC 6301 und PCC 7942 essentiellen Proteins IdiA. Thesis, Fakultät für Biologie. Bielefeld, Universität Bielefeld: 198.

Michel, K. P. (2003). Adaptation of the photosynthetic electron transport chain in cyanobacteria to iron deficiency and oxidative stress: The function of IdiA and IsiA. Fakultät für Biologie. Bielefeld, Bielefeld: 96.

Michel, K. P., S. Berry, et al. (2003). "Adaptation to iron deficiency: A comparison between the cyanobacterium *Synechococcus elongatus* PCC 7942 wild-type and a DpsA-free mutant." Photosynth Res 75: 71-84.

Michel, K. P., P. Exss-Sonne, et al. (1998). "Immunocytochemical localization of IdiA, a protein expressed under iron or manganese limitation in the mesophilic cyanobacterium *Synechococcus* PCC 6301 and the thermophilic cyanobacterium *Synechococcus elongatus*." Planta 205(1): 73-81.

Michel, K. P., F. Krüger, et al. (1999). "Molecular characterization of *idiA* and adjacent genes in the cyanobacteria *Synechococcus* sp. strains PCC 6301 and PCC 7942." Microbiology 145(Pt 6): 1473-1484.

Michel, K. P. and E. K. Pistorius (1992). "Isolation of a photosystem II associated 36 kDa polypeptide and an iron stress 34 kDa polypeptide from thylakoid membranes of the cyanobacterium *Synechococcus* PCC 6301 grown under mild iron deficiency." Z Naturforsch [C] 47: 867-874.

Michel, K. P. and E. K. Pistorius (2003). "Adaptation of the photosynthetic electron transport chain in cyanobacteria to iron deficiency: The function of IdiA and IsiA." Physiol Plant 119: 1-15.

Michel, K. P. and E. K. Pistorius (2004). "Adaptation of the photosynthetic electron transport chain in cyanobacteria to iron deficiency: The function of IdiA and IsiA." Physiol Plant 119: 1-15.

Michel, K. P., E. K. Pistorius, et al. (2001). "Unusual regulatory elements for iron deficiency induction of the *idiA* gene of *Synechococcus elongatus* PCC 7942." J Bacteriol 183(17): 5015-5024.

Michel, K. P., H. H. Thole, et al. (1996). "IdiA, a 34 kDa protein in the cyanobacteria *Synechococcus* sp. strains PCC 6301 and PCC 7942, is required for growth under iron and manganese limitations." Microbiology 142(Pt 9): 2635-2645.

Michel, K. P., N. Yousef, et al. (2003). Modification of photosystem II by IdiA and photosystem I by IsiA in *Synechococcus elongatus* PCC 7942 under iron limitation and regulation of these modifications. Functional Genomics in Cyanobacteria, Benediktbeuren, Germany.

Mittler, R. (2002). "Oxidative stress, antioxidants and stress tolerance." Trends Plant Sci 7(9): 405-410.

Mittler, R. and E. Tel-Or (1991). "Oxidative stress responses and shock proteins in the unicellular cyanobacterium *Synechococcus* R2 (PCC-7942)." Arch Microbiol 155: 125-130.

Mittler, R. and E. Tel-Or (1991a). "Oxidative stress responses in the unicellular cyanobacterium *Synechococcus* PCC 7942." Free Radic Res Commun 12-13(Pt 2): 845-850.

Moezelaar, R., S. M. Bijvank, et al. (1996). "Fermentation and sulfur reduction in the mat-building cyanobacterium *Microcoleus chthonoplastes*." Appl Environ Microbiol 62(5): 1752-1758.

Moezelaar, R. and L. J. Stal (1994). "Fermentation in the unicellular cyanobacterium *Microcystis* PCC 7806." Arch Microbiol 162: 63-69.

Mongkolsuk, S. and J. D. Helmann (2002). "Regulation of inducible peroxide stress responses." Mol Microbiol 45(1): 9-15.

Morgenstern, B. (1999). "DIALIGN 2: improvement of the segment-to-segment approach to multiple sequence alignment." Bioinformatics 15(3): 211-218.

Murata, N. (2003). Stress-induced gene expression and stress sensors in Synechocystis sp. PCC 6803. Functional Genomics in Cyanobacteria, Benediktbeuren, Germany.

Nachin, L., L. Loiseau, et al. (2003). "SufC: An unorthodox cytoplasmic ABC/ATPase required for [Fe-S] biogenesis under oxidative stress." Embo J 22(3): 427-437.

Nakamura, Y., T. Kaneko, et al. (2003). "Complete genome structure of Gloeobacter violaceus PCC 7421, a cyanobacterium that lacks thylakoids (supplement)." DNA Res 10(4): 181-201.

Neilands, J. B. (1995). "Siderophores: structure and function of microbial iron transport compounds." J Biol Chem 270(45): 26723-26726.

Neuhaus, H. E. and M. J. Emes (2000). "Nonphotosynthetic Metabolism in Plastids." Annu Rev Plant Physiol Plant Mol Biol 51: 111-140.

Newberry, K. J. and R. G. Brennan (2004). "The structural mechanism for transcription activation by MerR family member multidrug transporter activation, N terminus." J Biol Chem 279(19): 20356-62.

Nicholson, M. L. and D. E. Laudenbach (1995). "Genes encoded on a cyanobacterial plasmid are transcriptionally regulated by sulfur availability and CysR." J Bacteriol 177(8): 2143-2150.

Nield, J., E. P. Morris, et al. (2003). "Structural analysis of the photosystem I supercomplex of cyanobacteria induced by iron deficiency." Biochemistry 42(11): 3180-3188.

Nodop, A., D. Pietsch, et al. (2008). "Transcript profiling reveals new insights into the acclimation of Synechococcus elongatus PCC 7942 to iron starvation." Plant Physiol.147: 1-7.

Nodop, A., I. Suzuki, et al. (2006). "Physiological and molecular characterization of a Synechocystis sp. PCC 6803 mutant lacking histidine kinase Slr1759 and response regulator Slr1760." Z Naturforsch [C] 61(11-12): 865-78.

Nowalk, A. J., S. B. Tencza, et al. (1994). "Coordination of iron by the ferric iron-binding protein of pathogenic Neisseria is homologous to the transferrins." Biochemistry 33(43): 12769-12775.

Odom, W. R., R. Hodges, et al. (1993). "Characterization of Synechocystis sp. PCC 6803 in iron-supplied and iron-deficient media." Plant Mol Biol 23(6): 1255-1264.

Ogawa, T. (1991a). "A gene homologous to the subunit-2 gene of NADH dehydrogenase is essential to inorganic carbon transport of Synechocystis PCC6803." Proc Natl Acad Sci U S A 88(10): 4275-4279.

Ogawa, T. (1991b). "Cloning and inactivation of a gene essential to inorganic carbon transport in Synechocystis PCC 6803." Plant Physiol 96: 280-284.

Ogawa, T. and A. Kaplan (1987). "The stoichiometry between CO_2 and H^+ fluxes involved in the transport of inorganic carbon in cyanobacteria." Plant Physiol 83: 888-891.

Ohkawa, H., H. B. Pakrasi, et al. (2000). "Two types of functionally distinct NAD(P)H dehydrogenases in Synechocystis sp. strain PCC6803." J Biol Chem 275(41): 31630-31634.

Ohkawa, H., H. B. Pakrasi, et al. (2000a). "Two types of functionally distinct NAD(P)H dehydrogenases in Synechocystis sp. strain PCC 6803." J Biol Chem 275(41): 31630-31634.

Ohkawa, H., G. D. Price, et al. (2000b). "Mutation of ndh genes leads to inhibition of CO(2) uptake rather than HCO(3)(-) uptake in Synechocystis sp. strain PCC 6803." J Bacteriol 182(9): 2591-2596.

Ohkawa, H., M. Sonoda, et al. (2002). "Functionally distinct NAD(P)H dehydrogenases and their membrane localization in Synechocystis sp. PCC 6803." Funct Plant Biol 29: 195-200.

Ohkawa, H., M. Sonoda, et al. (2001). "Localization of NAD(P)H dehydrogenase in the cyanobacterium *Synechocystis* sp. strain PCC 6803." J Bacteriol 183(16): 4938-4939.

Omata, T. and N. Murata (1984). "Isolation and characterization of three types of membranes from the cyanobacterium *Synechocystis* PCC 6714." Arch Microbiol 139: 113-116.

Öquist, G. (1971). "Changes in pigment composition and photosynthesis induced by iron deficiency in the blue-green algae *Anacystis nidulans*." Physiol Plant 25: 188-191.

Öquist, G. (1974). "Iron deficiency in the blue-green algae *Anacystis nidulans*." Physiol Plant 30: 30-37.

Öquist, G. (1974). "Iron deficiency in the blue-green algae *Anacystis nidulans*: Fluorescence and absorption spectra recorded at 77 K." Physiol Plant 31: 55-58.

Outten, C. E., F. W. Outten, et al. (1999). "DNA distortion mechanism for transcriptional activation by ZntR, a Zn(II)-responsive MerR homologue in Escherichia coli." J Biol Chem 274(53): 37517-24.

Paerl, H. W. (2000). Marine plankton. The Ecology of Cyanobacteria: Their Diversity in Time and Space. B. A. Whitton and M. Potts. Dordrecht, London, Boston, Kluwer Academic Publishers: 121-148.

Panina, E. M., A. A. Mironov, et al. (2001). "Comparative analysis of FUR regulons in gamma-proteobacteria." Nucleic Acids Res 29(24): 5195-5206.

Pietsch, D. (2004). "Isolierung und teilweise characterisierung eine putativen Eisen-Schwefel-Proteins in dem cyanobacterium *Synechococcus elongatus* PCC 7942."

Pietsch, D., D. Staiger, et al. (2007). "Characterization of the putative iron sulfur protein IdiC (ORF5) in *Synechococcus elongatus* PCC 7942." Photosynth Res 94(1): 91-108.

Pistorius, E. K., K. Jetschmann, et al. (1979). "The dark respiration of Anacystis nidulans. Production of HCN from histidine and oxidation of basic amino acids." Biochim Biophys Acta 585(4): 630-42.

Pistorius, E. K. and H. Voss (1980). "Some properties of a basic L-amino acid oxidase from *Anacystis nidulans*." Biochim Biophys Acta 611: 227-240.

Prommeenate, P., A. M. Lennon, et al. (2004). "Subunit composition of NDH-1 complexes of *Synechocystis* sp. PCC 6803: identification of two new *ndh* gene products with nuclear-encoded homologues in the chloroplast Ndh complex." J Biol Chem 279(27): 28165-28173.

Ratledge, C. and L. G. Dover (2000). "Iron metabolism in pathogenic bacteria." Annu Rev Microbiol 54: 881-941.

Reddy, K. J., G. S. Bullerjahn, et al. (1988). "Cloning, nucleotide sequence, and mutagenesis of a gene (*irpA*) involved in iron-deficient growth of the cyanobacterium *Synechococcus* sp. strain PCC 7942." J Bacteriol 170(10): 4466-4476.

Riethman, H. C. and D. Sherman (1988). "Immunological characterization of iron-regulated membrane proteins in the cyanobacterium *Anacystis nidulans* R2." Plant Physiol 88: 497-505.

Riethman, H. C. and L. A. Sherman (1988). "Purification and characterization of an iron stress-induced chlorophyll- protein from the cyanobacterium *Anacystis nidulans* R2." Biochim Biophys Acta 935(2): 141-151.

Riethman, H. C. and L. A. Sherman (1988). "Regulation of cyanobacterial pigment-protein composition and organization by environmental factors." Photosynth Res 18: 133-161.

Rippka, R. (1988). "Isolation and Purification of Cyanobacteria." Methods Enzymol 167: 3-27.

Rowell, P., J. Diez, et al. (1981). "Molecular heterogeneity of ferredoxin:NADP+ oxidoreductase from the cyanobacterium Anabaena cylindrica." Biochim Biophys Acta 657(2): 507-16.

Rueckert, C., A. Puhler, et al. (2003). "Genome-wide analysis of the L-methionine biosynthetic pathway in *Corynebacterium glutamicum* by targeted gene deletion and homologous complementation." J Biotechnol 104(1-3): 213-228.

Rutherford, A. W. and A. Krieger-Liszkay (2001). "Herbicide-induced oxidative stress in photosystem II." Trends Biochem Sci 26(11): 648-653.

Samartzidou, H. and W. R. Widger (1998). "Transcriptional and posttranscriptional control of mRNA from *lrtA*, a light-repressed transcript in *Synechococcus* sp. PCC 7002." Plant Physiol 117(1): 225-234.

Sambrook, J., E. F. Fritsch, et al. (1989). Molecular Cloning: A Laboratory Manual. New York, Cold Spring Harbor Laboratory.

Samson, G., S. K. Herbert, et al. (1994). "Acclimation of the photosynthetic apparatus to growth irradiance in a mutant strain of *Synechococcus* lacking iron superoxide dismutase." Plant Physiol 105(1): 287-294.

Samuilov, V. D. (1997). "Photosynthetic oxygen: the role of H_2O_2. A review." Biochemistry (Moscow) 62(5): 451-454.

Sanders, J. D., L. D. Cope, et al. (1994). "Identification of a locus involved in the utilization of iron by *Haemophilus influenzae*." Infect Immun 62(10): 4515-4525.

Sandmann, G. (1985). "Consequences of iron deficiency on photosynthetic and respiratory electron transport in blue-green algae." Photosynth Res 6: 261-271.

Sandström, S., A. G. Ivanov, et al. (2002). "Iron stress responses in the cyanobacterium *Synechococcus* sp. PCC7942." Physiol Plant 116(2): 255-263.

Sandström, S., Y. I. Park, et al. (2001). "CP43', the *isiA* gene product, functions as an excitation energy dissipater in the cyanobacterium *Synechococcus* sp. PCC 7942." Photochem Photobiol 74(3): 431-437.

Schaegger, H. and G. von Jagow (1987). "Tricine-sodium dodecylsulphate-polyacrylamide gel electrophoresis for the separation of proteins in the range from 1 to 100 kDa." Anal Biochem 166: 368-379.

Scherer, S., I. Alpes, et al. (1988). "Ferredoxin-NADP+ oxidoreductase is the respiratory NADPH dehydrogenase of the cyanobacterium Anabaena variabilis." Arch Biochem Biophys 267(1): 228-35.

Schluchter, W. M. and D. A. Bryant (1992). "Molecular characterization of ferredoxin-NADP+ oxidoreductase in cyanobacteria: cloning and sequence of the petH gene of Synechococcus sp. PCC 7002 and studies on the gene product." Biochemistry 31(12): 3092-102.

Schmetterer, G. (1994). Cyanobacterial respiration. The Molecular Biology of Cyanobacteria. D. A. Bryant. Dordrecht, Boston, London, Kluwer Academic Publishers, The Netherlands. **4:** 409-435.

Schopf, J. W. (2000). The fossil record: Tracing the roots of the cyanobacterial lineage. The Ecology of Cyanobacteria: Their Diversity in Time and Space. B. A. Whitton and M. Potts. Dordrecht, London, Boston, Kluwer Academic Publishers**:** 13-35.

Schreiber, U., C. Klughammer, et al. (1988). "Measuring P700 absorbance changes around 830 nm with a new type of pulse modulation system." Z Naturforsch 43: 686-698.

Schriek, S. (2008). "Interrelationship of photosynthesis, respiration, and L-Arginin metabolism in the cyanobacterium *Synechocystis* sp. PCC 6803."

Sharma, R. C. and R. T. Schimke (1996). "Preparation of electrocompetent E. coli using salt-free growth medium." Biotechniques 20(1): 42-4.

Sherman, D. M. and L. A. Sherman (1983). "Effect of iron deficiency and iron restoration on ultrastructure of *Anacystis nidulans*." J Bacteriol 156(1): 393-401.

Sherman, L. A., P. C. Meunier, et al. (1998). "Diurnal rhythms in metabolism. A day in the life of a unicellular, diazotrophic cyanobacterium." Photosynth Res 58: 25-42.

Shibata, M., H. Ohkawa, et al. (2001). "Distinct constitutive and low-CO2-induced CO2 uptake systems in cyanobacteria: genes involved and their phylogenetic relationship with homologous genes in other organisms." Proc Natl Acad Sci U S A 98(20): 11789-94.

Simon, R. D. (1987). Inclusion bodies in the cyanobacteria: Cyanophycin, polyphosphate, polyhedral bodies. The Cyanobacteria. P. Fay and C. van Baalen. Amsterdam, New York, Oxford, Elsevier. 1: 199-225.

Singh, A. K., H. Li, et al. (2003). "Microarray analysis and redox control of gene expression in the cyanobacterium *Synechocystis* sp. PCC 6803." Physiol Plant: in press.

Singh, A. K., H. Li, et al. (2004). "Microarray analysis and redox control of gene expression in the cyanobacterium *Synechocystis* sp. PCC 6803." Physiol Plant 120(1): 27-35.

Singh, A. K. and L. A. Sherman (2006). "Iron-independent dynamics of IsiA production during the transition to stationary phase in the cyanobacterium *Synechocystis* sp. PCC 6803." FEMS Microbiol Lett 256: 159-164.

Smillie, R. M. (1965). "Isolation of phytoflavin, a flavoprotein with chloroplast ferredoxin activity." Plant Physiol 40: 1124-1128.

Smith, P. K., R. I. Krohn, et al. (1985). "Measurement of protein using bicinchoninic acid." Anal Biochem 150: 368-379.

Staudenmaier, H., B. Van Hove, et al. (1989). "Nucleotide sequences of the fecBCDE genes and locations of the proteins suggest a periplasmic-binding-protein-dependent transport mechanism for iron(III) dicitrate in Escherichia coli." J Bacteriol 171(5): 2626-33.

Stearman, R., D. S. Yuan, et al. (1996). "A permease-oxidase complex involved in high-affinity iron uptake in yeast." Science 271(5255): 1552-1557.

Stephan, D. P., H. G. Ruppel, et al. (2000). "Interrelation between cyanophycin synthesis, L-arginine catabolism and photosynthesis in the cyanobacterium *Synechocystis* sp. strain PCC 6803." Z Naturforsch [C] 55(11-12): 927-942.

Steunou, A. S., D. Bhaya, et al. (2006). "*In situ* analysis of nitrogen fixation and metabolic switching in unicellular thermophilic cyanobacteria inhabiting hot spring microbial mats." Proc Natl Acad Sci U S A 103(7): 2398-2403.

Steunou, A. S., D. Bhaya, et al. (2006). "In situ analysis of nitrogen fixation and metabolic switching in unicellular thermophilic cyanobacteria inhabiting hot spring microbial mats." Proc Natl Acad Sci U S A 103(7): 2398-403.

Stork, T., K. P. Michel, et al. (2005). "Bioinformatic analysis of the genomes of the cyanobacteria *Synechocystis* sp. PCC 6803 and *Synechococcus elongatus* PCC 7942 for the presence of peroxiredoxins and their transcript regulation under stress." J Exp Bot 56(422): 3193-3206.

Straus, N. A. (1994). Iron deprivation: Physiology and gene regulation. The Molecular Biology of Cyanobacteria. D. A. Bryant. Dordrecht, Boston & London, Kluwer Academic Publishers. 1: 731-750.

Sumegi, B. and P. A. Srere (1984). "Complex I binds several mitochondrial NAD-coupled dehydrogenases." J Biol Chem 259(24): 15040-15045.

Sweeney, W. V., A. J. Bearden, et al. (1974). "The electron paramagnetic resonance of oxidized clostridial ferredoxins." Biochem Biophys Res Commun 59(1): 188-94.

Tabita, F. R. (1987). Carbon dioxide fixation and its regulation in cyanobacteria. The Cyanobacteria. P. Fay and C. van Baalen. Amsterdam, New York, Oxford, Elsevier: 95-118.

Tabita, F. R. (1994). The biochemistry and molecular regulation of carbon dioxide metabolism in cyanobacteria. The Molecular Biology of Cyanobacteria. D. A. Bryant. Dordrecht, Boston & London, Kluwer Academic Publishers. 4: 437-467.

Tandeau de Marsac, N. and J. Houmard (1988). Complementary chromatic adaptation: Physiological conditions and action spectra. Methods Enzymol. L. Packer and A. N. Glazer, Academic Press. 167: 318-328.

Thomas, J. C., B. Ughy, et al. (2006). "A second isoform of the ferredoxin:NADP oxidoreductase generated by an in-frame initiation of translation." Proc Natl Acad Sci U S A 103(48): 18368-73.

Tichy, M. and W. Vermaas (1999). "*In vivo* role of catalase-peroxidase in *Synechocystis* sp. strain PCC 6803." J Bacteriol 181(6): 1875-1882.

Tindale, A. E., M. Mehrotra, et al. (2000). "Dual regulation of catecholate siderophore biosynthesis in *Azotobacter vinelandii* by iron and oxidative stress." Microbiology 146(Pt 7): 1617-1626.

Tölle, J., K. P. Michel, et al. (2002). "Localization and function of the IdiA homologue Slr1295 in the cyanobacterium *Synechocystis* sp. strain PCC 6803." Microbiology 148(Pt 10): 3293-3305.

Tortell, P. D., M. T. Maldonado, et al. (1999). "Marine bacteria and biogeochemical cycling of iron in the oceans." FEMS Microbiol Ecology 29: 1-11.

Trebst, A., B. Depka, et al. (2002). "A specific role for tocopherol and of chemical singlet oxygen quenchers in the maintenance of photosystem II structure and function in *Chlamydomonas reinhardtii*." FEBS Lett 516(1-3): 156-160.

Trick, C. G. and A. Kerry (1992). "Isolation and purification of siderophores produced by cyanobacteria, *Synechococcus* sp. PCC 7942 and *Anabaena variabilis* ATCC 29413." Curr Microbiol 24: 241-245.

Tucker, D. L., K. Hirsh, et al. (2001). "The manganese stabilizing protein (MSP) and the control of O_2 evolution in the unicellular, diazotrophic cyanobacterium, *Cyanothece* sp. ATCC 51142." Biochim Biophys Acta 1504(2-3): 409-422.

Unden, G. and J. R. Guest (1985). "Isolation and characterization of the Fnr protein, the transcriptional regulator of anaerobic electron transport in *Escherichia coli*." Eur J Biochem 146(1): 193-199.

van der Oost, J., B. A. Bulthuis, et al. (1989). "Fermentation metabolism of the unicellular cyanobacterium *Cyanothece* PCC 7822." Arch Microbiol 152: 415-419.

van Thor, J. J., O. W. Gruters, et al. (1999). "Localization and function of ferredoxin:NADP(+) reductase bound to the phycobilisomes of Synechocystis." Embo J 18(15): 4128-36.

van Thor, J. J., R. Jeanjean, et al. (2000). "Salt shock-inducible photosystem I cyclic electron transfer in *Synechocystis* PCC6803 relies on binding of ferredoxin:NADP(+) reductase to the thylakoid membranes via its CpcD phycobilisome-linker homologous N-terminal domain." Biochim Biophys Acta 1457(3): 129-144.

van Waasbergen, L. G., N. Dolganov, et al. (2002). "*nblS*, a gene involved in controlling photosynthesis-related gene expression during high light and nutrient stress in *Synechococcus elongatus* PCC 7942." J Bacteriol 184(9): 2481-2490.

Vermaas, W. F. J. (2001). Evolution of photosynthesis. Encyclopedia of Life Sciences. London, Nature Publishing Group. 1: 1-18.

Vermaas, W. F. J. (2001). Photosynthesis and respiration in cyanobacteria. Encyclopedia of Life Sciences. London, Nature Publishing Group. 1: 1-7.

Vinnemeier, J. and M. Hagemann (1999). "Identification of salt-regulated genes in the genome of the cyanobacterium *Synechocystis* sp. strain PCC 6803 by subtractive RNA hybridization." Arch Microbiol 172(6): 377-386.

Vinnemeier, J., A. Kunert, et al. (1998). "Transcriptional analysis of the *isiAB* operon in salt-stressed cells of the cyanobacterium *Synechocystis* sp. PCC 6803." FEMS Microbiol Lett 169(2): 323-330.

Von Wiren, N., S. Mori, et al. (1994). "Iron Inefficiency in Maize Mutant ys1 (Zea mays L. cv Yellow-Stripe) Is Caused by a Defect in Uptake of Iron Phytosiderophores." Plant Physiol 106(1): 71-77.

Wächtershäuser, G. (1990). "Evolution of the first metabolic cycles." PNAS 87(1): 200-204.

Wächtershäuser, G. (2000). "Origin of life: Life as we don't know it." Science 289(5483): 1307-1308.

Wang, T., G. Shen, et al. (2004). "The *sufR* gene (*sll0088* in *Synechocystis* sp. strain PCC 6803) functions as a repressor of the *sufBCDS* operon in iron-sulfur cluster biogenesis in cyanobacteria." J Bacteriol 186(4): 956-967.

Webb, R., T. Troyan, et al. (1994). "MapA, an iron-regulated, cytoplasmic membrane protein in the cyanobacterium *Synechococcus* sp. strain PCC 7942." J Bacteriol 176(16): 4906-4913.

Weidner, U., S. Geier, et al. (1993). "The gene locus of the proton-translocating NADH: ubiquinone oxidoreductase in Escherichia coli. Organization of the 14 genes and relationship between the derived proteins and subunits of mitochondrial complex I." J Mol Biol 233(1): 109-22.

Wenk, S. O. and J. Kruuip (2000). "Novel, rapid purification of the membrane photosystem I by high-performance liquid chromatography on porous materials." J Chrom B 737: 131-142.

Whitton, B. A. and M. Potts (2000). Introduction to the cyanobacteria. The Ecology of Cyanobacteria: Their Diversity in Time and Space. B. A. Whitton and M. Potts. Dordrecht, London, Boston, Kluwer Academic Publishers: 1-11.

Wydrzynski, T., J. Angström, et al. (1989). "H_2O_2 formation by photosystem II." Biochim Biophys Acta 973: 23-28.

Yagi, T. (1991). "Bacterial NADH-quinone oxidoreductases." J Bioenerg Biomembr 23: 211-225.

Yagi, T. (1993). "The bacterial energy-transducing NADH-quinone oxidoreductases." Biochem Biophys Acta 1141: 1-17.

Yagi, T. (1998). "Procaryotic complex I (NDH-1), an overview." Biochim Biophys Acta 1364: 125-133.

Yamanaka, G., A. N. Glazer, et al. (1978). "Cyanobacterial phycobilisomes. Characterization of the phycobilisomes of Synechococcus sp. 6301." J Biol Chem 253(22): 8303-10.

Yang, Y. H., S. Dudoit, et al. (2002). "Normalization for cDNA microarray data: a robust composite method addressing single and multiple slide systematic variation." Nucleic Acids Res 30(4): e15.

Yanisch-Perron, C., J. Vieira, et al. (1985). "Improved M13 phage cloning vectors and host strains: nucleotide sequences of the M13mp18 and pUC19 vectors." Gene 33(1): 103-19.

Yao, Y., T. Tamura, et al. (1984). "Spirulina ferredoxin-NADP+ reductase. The complete amino acid sequence." J Biochem 95(5): 1513-6.

Yeremenko, N., R. Jeanjean, et al. (2005). "Open reading frame *ssr2016* is required for antimycin A-sensitive photosystem I-driven cyclic electron flow in the cyanobacterium *Synechocystis* sp. PCC 6803." Plant Cell Physiol 46(8): 1433-1436.

Yeremenko, N., R. Kouril, et al. (2004). "Supramolecular organization and dual function of the IsiA chlorophyll-binding protein in cyanobacteria." Biochemistry 43(32): 10308-13.

Yousef, N., E. K. Pistorius, et al. (2003). "Comparative analysis of *idiA* and *isiA* transcription under iron starvation and oxidative stress in *Synechococcus elongatus* PCC 7942 wild type and selected mutants." Arch Microbiol 180: 471-483.

Zak, E., B. Norling, et al. (2001). "The initial steps of biogenesis of cyanobacterial photosystems occur in plasma membranes." Proc Natl Acad Sci U S A 98(23): 13443-8.

Zhang, H., J. P. Whitelegge, et al. (2001). "Ferredoxin:NADP$^+$ oxidoreductase is a subunit of the chloroplast cytochrome b_6f complex." J Biol Chem 276(41): 38159-38165.

Zhang, P., N. Battchikova, et al. (2004). "Expression and functional roles of the two distinct NDH-1 complexes and the carbon acquisition complex NdhD3/NdhF3/CupA/Sll1735 in Synechocystis sp PCC 6803." Plant Cell 16(12): 3326-40.

Zheng, M., B. Doan, et al. (1999). "OxyR and SoxRS regulation of *fur*." J Bacteriol 181(15): 4639-4643.

8 Appendix

8.1 Supplementary tables for chapter 3.7

Supplementary table 1: List of differentially-regulated genes encoding regulatory proteins, cofactor and pigment biosynthesis-related proteins, nucleic acid metabolism-related proteins as well as proteins of unknown function or hypothetical proteins from *S. elongatus* PCC 7942 WT, the IdiB-free mutant K10, and the *idiC*-merodiploid mutant MuD in response to growth for 24 (WT) or 72 h (WT, K10, and MuD) with iron-deficient vs. iron-sufficient BG11 medium. The table contains the evaluated data of three biological and two technical replicates and includes a dye-swap experiment. The fold change value is calculated as $\log_2^{M\text{-value}}$ of M-values with corresponding p-values ≤0.051. M-values >-0.90 and <+0.90 indicate no significant change in the transcriptional level (corresponding to a fold change of ≤1.87 and ≥0.53). Significantly increased or decreased transcript levels are printed in bold letters. JGI ORFs correspond to the JGI annotation. Common gene names are given in the column to the right. "NA" means that the gene is not annotated in the JGI annotation system.

JGI ORF	Gene	Annotated protein function	Fold change			
			Growth -Fe vs. +Fe		Growth -Fe vs. +Fe	
			WT 24 h	WT 72 h	K10 72 h	MuD 72 h
Transcripts of genes encoding regulatory proteins						
1316	2731	Putative transcription factor, similar to Tlr1758 of *Thermosynechococcus elongatus* BP-1	1.45	**2.00**	1.20	1.17
1402	2777	Hypothetical 48 kDa protein with zinc-finger protein-protein-interaction domain and a histone deacetylase domain being typical for transcriptional regulators	1.04	0.85	0.66	**0.47**
2466	1044	CheY-like response regulator with winged-helix DNA-binding domain, similar to Rre37 (Sll1330 of *Synechocystis* sp. PCC 6803) and Slr0947 (Ycf27-like)	**2.08**	**2.22**	1.02	0.90
1784	sigF2	Group III RNA polymerase σ factor	1.03	0.72	0.59	**0.49**
0672	rpoD3	Alternative group II RNA polymerase σ factor	0.85	**2.03**	**2.38**	1.12
0569	rpoD4	Alternative group II RNA polymerase σ factor	**2.04**	**2.15**	1.29	0.60
2121	0652	Anti-σ F factor antagonist similar protein	1.57	**2.00**	**2.16**	1.66
2352	lrtA	Light-repressed protein A and putative σ^{54} factor-modulating ribosomal protein S30EA	0.69	**0.35**	**0.21**	**0.25**
Transcripts of genes encoding cofactor- and pigment biosynthesis-related proteins						
0597	menB	Dihydroxynaphthoic acid synthetase or naphthoate synthase EC:4.1.3.36, menaquinone biosynthesis	**2.03**	1.73	1.61	0.90
1513	dxr	1-Deoxy-D-xylulose 5-phosphate reductoisomerase, in terpenoid orisoprenoid biosynthesis	0.75	0.74	0.85	**0.48**
1419	bchL	Light-independent protochlorophyllide reductase, [Fe-S] ATP-binding protein	1.22	**0.50**	0.66	**0.42**
1420	bchN	N subunit of light-independent protochlorophyllide reductase	1.08	**0.53**	0.67	**0.42**
2503	lpqR	Light-dependent protochlorophyllide reductase	0.95	**0.47**	**2.28**	0.58
1858	ho-1	Heme oxygenase	1.07	0.66	0.68	**0.40**
Transcripts of genes encoding nucleic acid metabolism-related proteins						
2117	gatA	Glutamyl-tRNA-amidotransferase subunit A	1.24	1.39	1.84	**2.81**

2119	spoU	tRNA-rRNA methyltransferase	1.28	1.68	**2.33**	**2.57**
2120	0651	Ribonuclease III domain-containing protein, inactive ribonuclease III homolog lacking charged active site amino acid residues	1.43	2.07	**3.05**	**3.36**
2207	truA	Pseudouridine synthase involved in tRNA processing	1.18	1.23	1.60	**2.01**
2209	rpoA	DNA-directed RNA polymerase	1.04	1.09	**2.30**	**2.30**

Transcripts of genes encoding protein biosynthesis-related proteins

2082	0609	FusA translation elongation factor EF	0.65	0.63	0.57	**0.23**
0885	fusA	FusA translation elongation factor G	1.01	1.13	**2.79**	**1.99**
2226	rpsC	Ribosomal protein S3	1.21	1.08	1.54	**2.40**
2219	rpsH	Ribosomal protein S8	1.34	1.20	**2.83**	**2.33**
2210	rps11	Ribosomal protein S11	0.95	0.97	**2.48**	**2.04**
2210	rpsM	Ribosomal protein S13	0.90	0.94	1.77	**2.06**
1123	rpsR	Ribosomal protein S18	1.24	1.16	1.41	**2.12**
2232	rplC	Ribosomal protein L3	1.27	0.95	1.32	**2.10**
2218	rpl6	Ribosomal protein L6	1.28	1.21	**2.33**	**2.01**
2206	rplM	Ribosomal protein L13	1.27	1.28	**2.70**	**2.13**
2222	rplN	Ribosomal protein L14	1.21	1.16	**2.01**	**2.13**
2225	rplP	Ribosomal protein L16	1.17	1.09	1.83	**2.30**
2208	rplQ	Ribosomal protein L17	1.10	1.13	**2.57**	**2.07**
2227	rplV	Ribosomal protein L22	1.15	1.03	1.45	**2.25**
2230	rplV	Ribosomal protein L23	1.24	0.99	1.88	**2.20**
2221	rplX	Ribosomal protein L24	1.20	1.20	**2.71**	**2.23**
2224	rpmC	Ribosomal protein L29	1.26	1.11	1.84	**2.51**
1122	rpmG	Ribosomal protein L33	1.22	1.06	1.45	**2.00**
1567	0047	Putative ribosomal protein L36	1.21	1.02	0.58	**0.41**

Transcripts of genes encoding proteins of different function

1606	0092	Growth factor-induced cell surface protein probably involved in cell adhesion processes, FAS1 domain-containing protein with fasciclin repeats	**2.10**	1.08	1.31	**2.32**
1870	hlyD	Secretion of proteins independently of ABC transporters	**1.95**	**2.04**	1.83	**2.40**
2450	pilG	Type IV pilus-assembly protein	1.57	**2.06**	1.40	1.11
2590	1185	Pilin-like hypothetical protein	1.07	0.82	**0.38**	**0.43**
NA	1227	Putative lipoprotein	**3.60**	**2.20**	**2.87**	1.43
0598	nlpD	Putative peptidoglycan peptidase	**2.30**	**3.32**	**2.64**	1.47
0779	2085	Hypothetical seven transmembrane domain extracellular protein	1.61	**2.25**	**1.99**	1.80

Transcripts of genes encoding hypothetical proteins with predicted domains

1642	0133	Hypothetical 38 kDa protein of a family of conserved proteins with unknown functions	1.60	1.42	1.74	**2.10**
1822	0333	Putative lipoprotein with unknown function	1.29	**2.10**	**3.89**	**2.16**
0197	1419	Integral membrane protein with unknown function	1.09	**2.00**	**2.59**	1.68
0415	1674	Hypothetical 35 kDa protein with pentapeptide-repeat protein with unknown function	0.69	**0.46**	**0.46**	0.83
0390	1678	Hypothetical 12.8 kDa protein of a family of conserved proteins with unknown functions	0.72	1.09	0.82	**0.50**
0515	1782	Hypothetical 15.1 kDa protein of a family of eubacterial and archaea-bacterial proteins,	1.25	1.38	1.77	**2.43**

0834	2146	contains the HXXXEXX(W/Y)-motif, which may be involved in metal binding. Hypothetical protein with a conserved cyanobacterial protein domain	0.97	1.23	0.96	**0.48**
1287	2656	Protein of unknown function with hemolysin-type Ca^{2+}-binding domains and four nonapeptide repeats	**2.28**	**2.87**	1.38	0.57

Transcripts of genes encoding hypothetical proteins without predictable domains

1661	0155	Hypothetical 19.5 kDa protein	0.89	0.77	**0.14**	**0.27**
1757	0262	Hypothetical 10.5 kDa protein	**2.19**	0.72	**0.20**	**0.13**
NA	0263	Hypothetical 6.4 kDa protein	1.83	0.68	**0.24**	**0.19**
1797	0305	Hypothetical 18.4 kDa protein	**4.00**	**4.72**	1.02	0.85
1845	0357	Hypothetical 8.6 kDa protein	0.64	**0.34**	0.52	**0.47**
2003	0521	Hypothetical 25.4 kDa protein	0.94	1.55	1.63	**2.16**
NA	0523	Hypothetical 3.6 kDa protein	**3.14**	**2.41**	0.79	1.17
2059	0585	Hypothetical 21 kDa protein	**1.84**	**2.17**	1.15	1.09
2169	0707	Hypothetical 11.4 kDa protein, gene lies in opposite orientation upstream of fur(2)	**2.57**	**2.28**	1.40	1.79
2178	0718	Hypothetical 11.8 kDa membrane protein	**2.45**	**2.46**	1.25	0.91
NA	0790	Hypothetical 7.6 kDa protein	1.13	**0.39**	**0.30**	**0.21**
2260	0806	Hypothetical 13.7 kDa protein	0.86	1.13	1.54	**2.16**
2259	0852	Hypothetical 40.5 kDa protein	0.93	1.03	n.d.	**2.19**
2307	0868	Hypothetical 38.5 kDa protein	0.93	0.97	0.77	**0.40**
NA	0911	Hypothetical 4.3 kDa protein, gene lies immediately upstream of psaL	0.90	0.47	**0.32**	**0.26**
NA	0974	Hypothetical 4.6 kDa protein	1.10	0.60	**0.51**	**0.44**
2419	orfG	Hypothetical protein	1.66	**2.87**	0.66	0.88
0013	1224	Hypothetical 25.8 kDa protein	1.43	1.08	1.06	**0.44**
0037	1248	Hypothetical 8 kDa protein	**2.14**	**2.66**	1.40	1.06
NA	1298	Hypothetical 6.8 kDa protein	1.41	0.80	**0.39**	**0.44**
NA	1425	Hypothetical 4 kDa protein	n.d.	**2.00**	1.36	n.d.
0364	1616	Hypothetical 19.5 kDa protein	1.27	**2.33**	**2.22**	1.42
0342	1589	Hypothetical 13 kDa protein, located immediately upstream of psb27	**2.27**	**2.08**	1.42	1.13
NA	2206	Hypothetical 5.3 kDa protein	1.02	1.02	**2.17**	**2.22**
NA	2810	Hypothetical 3.8 kDa protein	0.98	0.85	0.85	**0.47**
0465	1728	Hypothetical membrane protein	1.31	**2.55**	1.16	1.06
0229	1459	Hypothetical 20 kDa protein	1.15	1.02	1.89	**2.10**
0253	1487	Hypothetical 17.6 kDa protein	1.10	1.24	2.41	**2.38**
0259	1494	Hypothetical 11.4 kDa protein	0.74	0.62	0.33	**0.33**
NA	1502	Hypothetical 4.9 kDa protein	0.84	**0.52**	**0.31**	**0.26**
NA	1534	Hypothetical 4.2 kDa protein, gene locates immediately upstream of psbO	0.56	**0.50**	**0.50**	0.55
0373	1627	Hypothetical protein	0.99	1.06	1.50	**2.08**
0444	sek0026	Hypothetical protein	0.99	0.80	1.85	**2.04**
0452	sek0018	Hypothetical protein	0.70	**0.46**	0.57	0.65
0551	1824	Hypothetical 13 kDa protein	0.90	**0.49**	0.60	0.61
0979	2106	Hypothetical 9.3 kDa protein	0.57	**0.38**	**0.32**	**0.48**
NA	2365	Hypothetical 4.6 kDa protein	1.04	**2.11**	1.49	**1.85**
NA	2439	Hypothetical 3.9 kDa protein	1.13	**0.36**	**0.20**	**0.24**
1397	2772	Hypothetical 43.8 kDa protein	1.09	0.78	0.57	**0.36**

8.2 Abbreviations

%	Percent
°C	Degree Celsius
2#1	MerR-free *S. elongatus* PCC 7942 mutant
aa	Amino acid residue
AckA	Acetat kinase A
ADP	Adenosin-5´-diphosphate
Al	Aluminium
Amp	Ampicillin
AP	Alkaline phosphatase
APC	Allophycocyanin
APS	Ammoniumperoxidsulfate
Arg	Arginine
ATP	Adenosin-5`-triphosphate
AtpA	ATPase A
H_2O bidest.	Bidestilled water
Bisacrylamid	N,N-Methylen-bisacrylamide
BLAST	Basic Local Alignment Search Tool
BN	Blue native
bp	Basepair
BPB	Bromophenolblue
BSA	Bovine Serum Albumine
C	Carbon
Ca	Calcium
CA	Carboanhydrase
Chl	Chlorophyll
Cm	Chloramphenicol
CMF-PBS	Calcium-magnesium-free phosphate-buffered saline
Cmp	Cytoplasmic membrane transport
CO_2	carbon dioxide
CoA	Coenzyme A
CP43	Gene product of psbC
CP47	Gene product of psbB
CS	Cell suspension
CTAP	Cetyl trimethyl ammonium bromide

Cu	Copper
Cyt c	Cytochrome c
Cyt b_6/f	Cytochrome b_6/f-complex
D1	Gene product of psbA
D2	Gene product of psbD
DCMU	3-(3,4-Dichlorphenyl)-1,1-Dimethyle urea
DEPC	Diethylpyrocarbonate
DH	Dehydrogenase
Dig	Digoxigenin
DM	Dodecyl maltoside
DNA	Desoxyribonucleic acid
DNase	Desoxyribonuclease
dNTP	2´-Desoxy(ribo)-nucleic acid-triphosphate
ds	Double-stranded
DTE	Dithioerythritol
DTT	Dithiothreitole
dUTP	desoxy-ridinetriphosphate
E	Extinction
ECL	Electrochemical Luminiscence
EDTA	Ethylendiamintetraacetate
Epps	N-(2-Hydroxyethyl)-Piperazin-N'-(3-Propansulfonic acid); HEPPS
EPR	Electron paramagnetic resonance spectroscopy
EtBR	Ethidiumbromide
EtOH	Ethanol
ExPASy	Expert Protein Analysis System
FAD	Flavineadenindinucleotide
Fd	Ferredoxin
Fe	Iron
[Fe-S]	Iron-sulfur center
FMN	Flavinemononucleotide
FNR	Ferredoxin-NADP-Oxidoreductase
FPE	French Press Extract
Ftr	Ferredoxin:thioredoxin reductase
Fur	Ferric uptake repressor
Fut	Ferric uptake system
g	Gramme

Abbreviations

G	Gauss
G/C	guanidin/cytosine content
h	hour
H_2O	Water
H_2O_2	Hydrogen peroxide
HEPES	4-(2-Hydroxyethyl)-1-Piperazin-Ethansulfonic acid
HIC	Hydrophobic interaction chromatography
His	Histidine
HPLC	High-performance liquid chromatography
HRP	Horse reddish peroxidase
HUP	Uptake hydrogenase
Hz	Hertz
ICP-OES	Inductively coupled plasma optical emission spectroscopy
Idi	Iron deficiency induced protein
IEC	Ion exchange chromatography
IgG	Immunglobuline G
IPTG	Isopropyl β-D-1-thiogalactopyranoside
Irp	Iron regulated protein
Isi	Iron stress induced protein
K	Kelvin
K9#1	FurII-free *S. elongatus* PCC 7942 mutant
K10	IdiB-free *S. elongatus* PCC 7942 mutant
kDa	Kilodalton
Km	Kanamycin
L	Litre
L-aa	L-amino acid
L-Aox	L-Amino acid oxidase
L-AOX/DH	L-Amino acid oxidase/dehydrogenase
L-arg	L-arginine
LB	Luria-Bertani-Broth
L-glu	L-glutamate
µg	Microgramme
µL	Microlitre
µM	Micromolar
µm	Micrometer
µmol	Micromole

M	Molar
MALDI-TOF MS	Matrix assisted laser desorption/ionisation time of flight mass spectroscopy
MapA	Membrane associated protein A
MM	Molecular Mass
MerR	Mercury response regulator
mg	Milligramme
Mg	Magnesium
min	Minute
mL	Millilitre
mM	Millimolar
Mn	Manganese
mol	Mole
mRNA	messenger-RNA
MSP	Manganese-stabilising Peptid (PsbO)
MuD	IdiC-merodiploid *S. elongatus* PCC 7942 mutant
mV	Millivolt
N	Nitrogen
N_2	Molecular nitrogen
$NADP^+$	oxidized Nicotinamide-adenine-dinucleotide(-phosphate)
$NADPH+H^+$	reduced Nicotinamide-adenine-dinucleotide(-phosphate)
NC	Nitrocellulose
NBT	Nitro blue tetrazolium
NCBI	National Center for Biotechnological Information
NDH-1	NADH-Dehydrogenase (Typ 1)
NDH-2	NADH-Dehydrogenase (Typ 2)
NH_4^+	Ammonia
Ni	Nickel
Ni-NTA	Ni-Nitrilotriacetic acid
NIR	Ferredoxin:nitrite reductase
nm	Nanometer
nM	Nanomolar
nt	Nucleotide(s)
NtcA	Global nitrogen regulator
Nuo	Substrate-binding subunit of the bacterial NDH-1 of *E. coli*
O_2	Molecular oxygen
$O_2^{\cdot -}$	Superoxid anion

Abbreviations

$^1O_2$	Singulet oxygen
OD_{xyz}	Optical density at wave length xyz nm
OH^-	Hydroxyl radical
OPR	One-Phor-All
ORF	Open-reading frame
P_i	Inorganic phosphate
P_{680}	Pigment 680, reaction center chlorophyll of PS II
P_{700}	Pigment 700, reaction center chlorophyll of PS I
p.a.	pro analysis
PAGE	Polyacrylamide gel electrophoresis
PAM	Puls amplitude modulation
PBS	Phycobilisome
PC	Plastocyanin
PCC	Pasteur Culture Collection
PCR	Polymerase Chain Reaction
pcv	packed cell volume
PetA	Rieske protein
PGA	Phospho clyceric acid
Pgam	Phosphoclycerate mutase
pI	Isoelectric point
pmol	Picomol
PphA	Phosphatase of the PP2C family
PQ	Plastoquinone
PS	Photosystem
PsaAB	Main subunit of PS I
PsbO	Manganese-stabilizing peptide (MSP)
Q	Chinone
xyz^R	Antibiotica resistance against xyz
RNA	Ribonucleic acid
RNase	Ribonuclease
Rpm	Rounds per minit
ROS	reactive oxygen species
RT	Room temperature
RT-PCR	Reverse-transcribed PCR
s	Second
S	Sulfur

Sbt	Sodium-bicarbonate transport
S-DH	Substrate-dehydrogenase
SDS	Sodium Dodecylsulfate
ss	Single stranded
sp.	Species
Sp	Spectinomycin
SQR	Sulfoquinosyl dehydrogenase
SU	Subunit
Suf	[Fe-S] assembly system
TBS	Tris-buffered NaCl-solution
TCC	Tricarboxylic acid cyclus
TEMED	N,N,N',N'-tetramethylethylendiamine
Tm	Melting temperature of a primer
Tricine	N-[Tris(hydroxyl methyl)-methyl]-glycine
Tris	Tris-(hxydroxymethyl)-aminomethan
TRX	Thioredoxin
TWEEN 20	Polyoxyethylensorbitanmonolaureate
U	Enzyme unit
UV	Ultra violet
v	Volume
V	Volt
v/v	volume per volume
W	Watt
WT	Wild type
w/v	weight per volume
x g	Fold gravity
Y	Yield
Z48754	EMBL nucleotide database entry for a 5.8 kb *Hin*dIII DNA-fragment of *Synechococcus elongatus* PCC 6301
Zn	Zin

Die VDM Verlagsservicegesellschaft sucht für wissenschaftliche Verlage abgeschlossene und herausragende

Dissertationen, Habilitationen, Diplomarbeiten, Master Theses, Magisterarbeiten usw.

für die kostenlose Publikation als Fachbuch.

Sie verfügen über eine Arbeit, die hohen inhaltlichen und formalen Ansprüchen genügt, und haben Interesse an einer honorarvergüteten Publikation?

Dann senden Sie bitte erste Informationen über sich und Ihre Arbeit per Email an *info@vdm-vsg.de*.

Sie erhalten kurzfristig unser Feedback!

VDM Verlagsservicegesellschaft mbH
Dudweiler Landstr. 99   Telefon +49 681 3720 174
D - 66123 Saarbrücken   Fax   +49 681 3720 1749
www.vdm-vsg.de

Die VDM Verlagsservicegesellschaft mbH vertritt

Printed by Books on Demand GmbH, Norderstedt / Germany